Method Quarterly

ISSUE 1 | BOUNDARIES

MARS SURFACE, NASA IMAGE LIBRARY

I0502861

SCIENCE

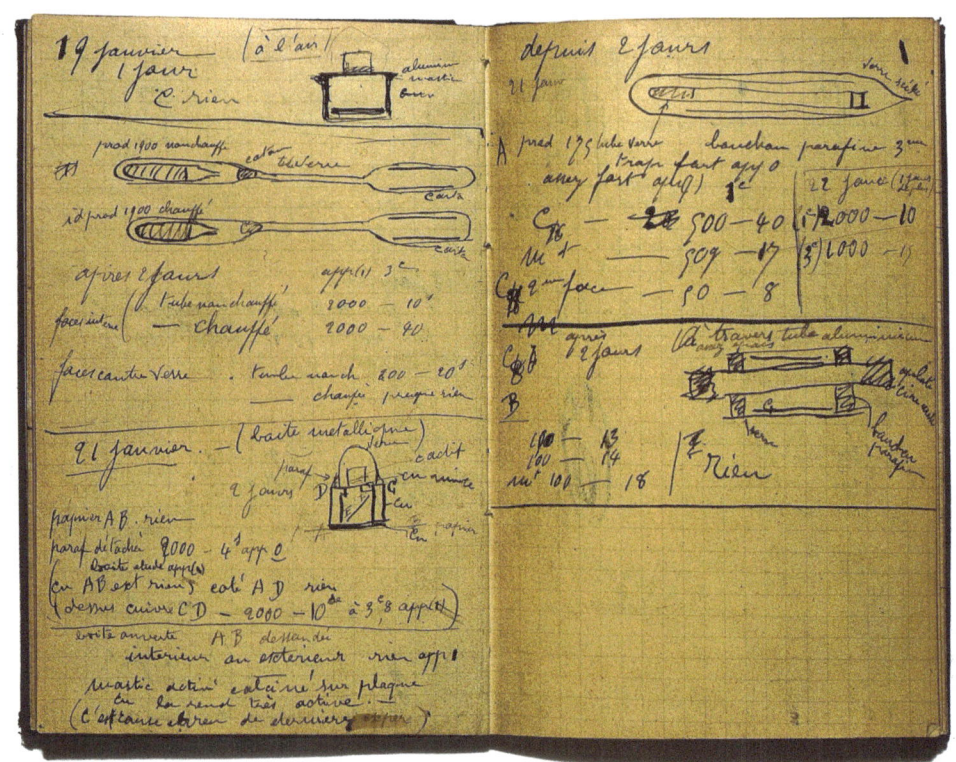

MARIE CURIE'S LAB NOTEBOOK, WELLCOME LIBRARY

in the making

As a scientist, Evelyn Fox Keller hated experiments. Today a professor emerita in the History of Science at MIT, she started out her scientific career in theoretical physics before jumping to molecular biology. What got her was the trivial volatility inherent in experimentation. "Someone could brush by the thermostat on the water bath, and a months' work would be down the drain. You didn't have control over it."

The methods of theoretical physics better suited her. "Just by thinking, you could deduce the way the world is. I could think about questions like, *Why is space three dimensional*? There were kinds of answers you could come up with, just with thinking in theoretical formal terms...But it was all very other-worldly; it had nothing to do with the concrete world in which we live. I was totally intoxicated by these ideas, thinking worlds into being, out of pure thought."

Both theoretical physics and molecular biology are scientific disciplines, but what does it mean that their routes to knowledge can be so different?

While the scientific method is routinely described as a linear path from observation to hypothesis to experiment to discovery, the everyday of scientific labor is much less ordered. It's a process that's circuitous, iterative, and often serendipitous. It can involve transcendent experiences of thinking the world into being alongside random accidents and detours. As a result, the realities of science in the making are simultaneously much more mundane—and much more interesting—than what the idealized version of the scientific method might lead us to believe.

These ideals also constrain how we define, understand, and talk about science. Science is bounded on all sides by what it isn't, so often defined in opposition to the messy world of human actions. Science is defined against art,

politics, religion, and culture. We're told it's not emotional, irrational, mystical, or personal. At the same time, science is criss-crossed by the internal boundaries splintering disciplines, isolating people based on methods, goals, and language.

In creating *Method Quarterly*, we wanted to bring together scientists, scholars, artists, and writers across the many fractured boundaries of science, to explore the rough edges and fuzzy uncertainties of scientific methods. Our first issue features pieces by academic and industry scientists, by an anthropologist studying science and "other" knowledges, a policy researcher, a poet, and a professor of experimental fiction, and of course, amazing science writers. They feature stories of people, places, and things that remain unbounded, unable to fit into any one category.

With "Boundaries," we hope to explore the grey areas in and around the edges of science. Doing so might help us better understand what counts as "science" in the first place (and what doesn't), who makes it (and who doesn't), and how we might go about subverting or breaking down those boundaries altogether.

We explore the vital imperatives of nuclear containment, juxtaposed with its mundane realities (with a side of kitty litter). We participate in a Ganzfeld experiment, which measures psychic abilities, to look at what the blurry boundaries between science and pseudoscience tell us about scientific methods. We look at the lines—man-made and feeble— that separate nature and the sprawling untame of concrete that is Los Angeles. And we learn from Evelyn Fox Keller's disciplinary border crossings and how she transformed concepts of gender in science. And much, much more.

We're very excited to share these stories with you, and we hope you'll stick around as we keep asking, *what is this thing called science*? ■

Janny Li

I Don't Have ESP

I begin my Ganzfeld experiment in a small soundproof booth. Dr. Nancy Sondow, the president of the American Society for Psychical Research (ASPR), has taped two translucent ping pong ball halves over my eyes to block visual stimuli. She had also placed over my ears a pair of noise canceling headphones continuously looping white noise to block auditory stimuli. A red floodlight aimed at my face—creating an undiffused pattern of light—completes the full sensory deprivation experience. Following fifteen minutes of guided breathing and relaxation exercises, I am ready for telepathic communication.

As an anthropologist interested in the history of scientific experiments, I study how "science" is and has always been defined against an "other." The Ganzfeld experiment used by parapsychologists working at the fringes of science can help us understand scientific practices and experiments in other contexts. For instance, the Ganzfeld experiment can tell us about how scientific novelties are often produced by forces—uncertainty, unforeseen contingencies, and as we will later see, surprises—that lie outside of the scientific method. And, it encourages us to rethink experiments not simply as generators of scientific knowledge, but also as provisional relationships that enable researchers access to their otherwise inaccessible objects of inquiry, be they psychic phenomena or elementary particles.

Two floors below the sensory deprivation chamber in the ASPR library, Patrice Keane, the society's executive director, randomly chooses one out of

Janny Li is a doctoral candidate in anthropology at the University of California, Irvine. Her research explores the boundaries between science and "other" knowledges, such as religion and the occult, and their broader implications for understanding scientific authority and knowledge four images. In this double-blind experiment, she is the "sender," focusing on an image in an attempt to target it to me in the soundproof booth upstairs. Sensory deprivation is thought to create the optimal conditions for psychic experiences and more practically, it safeguards against potential information leakage. I am told to wish for my target image. It should effortlessly appear in my mind sometime during the sensory deprivation session.

The Ganzfeld experiment is designed to measure what Sondow calls "psi" or exceptional mental states. Psi encompasses forms of extrasensory perception or anomalous mind-matter interactions that include: telepathy (direct mind-to-mind interaction), precognition, (perceiving things or events in the future), and clairvoyance (knowing at a distance). The experiment gained traction at the ASPR in the 1970s as an efficient shortcut to the more costly research on dream telepathy. The ASPR currently uses the Ganzfeld experiment in their attempts to correlate telepathy with diverse variables, such as personality types, mood, gender, age, and physiological reactions (e.g., ovulation, hormone levels).

The white noise resumes. It sounds like a whirling engine. I imagine myself hurling through outer space in a shuttle. Sondow's disembodied voice over my headphones instructs me to "think out loud." I list out loud the images that randomly appear to me: stars, black sky, blue and purple rolling mist, metal beams, mountains, pine trees, a lake, a teapot. I look out of the window of the shuttle and see a black sky blanketed by glowing stars. As I round a corner, I can see the sun slightly eclipsed by a planet with a slim crescent brightly peeking through. I keep repeating to Sondow, "it feels like I am in outer space." Flashes of my cat keep appearing to me, but I am too afraid to say this out loud, unsure of the mixing between my own imagination with bits and pieces of the target image. The session ends after thirty minutes.

As an experimental subject, I am what Sondow considers an "ordinary person," that is, someone who does not self-identify as possessing extrasensory perception. In the course of her research, Sondow has been surprised to find that many "ordinary" people have positive results in her Ganzfeld experiments. In a 1994 interview, Sondow discussed her belief in the pervasiveness of psi ability: "I [came] to the conclusion that psi is not a rare or extraordinary phenomenon, but

is probably a ubiquitous—but unconscious—phenomena that goes on all the time. It's a matter of interpretation, what some call extraordinary coincidences, others label psi." For Sondow, the task of finding psi phenomena among ordinary people occurs amidst larger uncertainties over how one interprets the nature of psi (e.g., origins and mechanics) and its existence more generally.

My particular Ganzfeld experiment was part of Sondow's recent efforts to correlate psi with levels of creativity. Outside of the sensory deprivation chamber, I fill out a series of surveys and questionnaires assessing my creativity, gauged by my dreaming patterns, open-mindedness, sense of well being, ability to adapt to change, and overall mood. The surveys asked me to rate my overall ease of achieving altered states of consciousness as well as the power of my imagination to generate and interpret visual imagery. Based upon my own assessments, Sondow hypothesized that I was a likely candidate to be predisposed and receptive to telepathic communication.

In Sondow's office, I receive a transcript of my observations and a set of four images: A) a green cartoon-like jungle scene with lush tress and vines, B) a red portrait of a king, C) a black scene of outer space, and D) a yellow impressionist painting of a castle. As I glance at the images I am immediately surprised. Image C is nearly identical to what I had visualized during my sensory deprivation session, complete with a black sky blanketed by stars, a blue and purple ringed planet in the center of the image, the metal beams of a satellite in the lower left-hand corner, and most surprisingly, a cat tumbling on the right. Shocked by the uncanny resemblance, I quickly blurted out to Sondow: "That's the target image, I'm positive!"

Despite my certainty, Sondow directs my attention to a prepared handout. She insists that I categorize each image from my transcript (e.g., teapot, stars, pine trees) into columns labeled A through D representing each of the four possible target images. She then asks me to assign a percentage to each column indicating my level of certainty regarding its status as the target image. I promptly assign the value of eighty percept under column C. Following my lead, Sondow completes her own evaluations and also assigns image C a value of eighty percent.

A sealed envelope containing the target image waits for us at the library downstairs. On the way,

Sondow explains to me that the Ganzfeld experiment is a double-blind test and that only the sender knows the identity of the target image. Sondow and I remain oblivious.. The double blind procedure is important because it ensures that Sondow does not in any way influence my evaluations. We open the envelope and I'm surprised again: I had been so sure that the image was C, but inside the envelope was image B, the portrait of a king. Despite its eerie coincidences, my Ganzfeld experiment is a failure. I didn't correctly identify the target image. Turns out, I don't have ESP.

The success of a Ganzfeld experiment, for Sondow, is predicated solely upon her subjects correctly identifying the target image. From an experiential standpoint, however, the success or failure of my test is not so rigidly determined. In fact, it raised some interesting questions for both Sondow and myself. While I did not correctly identify the target image, I did experience visions that remarkably corresponded to one of the four possible target images. If psi is potentially unconscious, Sondow offered as a tentative explanation, then it could be possible that Patrice consciously chose the image of a king, but unconsciously projected impressions of the other images. Whether or not Sondow and I agree or disagree over the results of my Ganzfeld experiment or whether or not I choose to interpret my experiences as psi or simply a extraordinary accident, my results open a space for Sondow to question other possible trajectories of telepathic communication and to develop alternate understandings of psi.

Parapsychologists conduct their Ganzfeld experiments within a milieu of uncertainty. They don't possess a working knowledge of the origins or mechanics of psi and—perhaps more significantly—they have yet to establish its existence within the scientific community and the public writ large. Given these murky conditions, parapsychologists often encounter surprises throughout the design, implementation, and analysis process as they attempt to refine their Ganzfeld experiments.

"Surprise" is not often a word associated with science, though other words that come to mind including "epiphanies," "discoveries," and "breakthroughs," are often used to describe "eureka" moments. This could be due to the fact that we most often associate science with rationality, logic, or deductive reasoning that at the outset precludes any notions of serendipity. However, as Hans-Jorg Rheinberger argues in *Toward a History of Epistemic Things*, scientific novelties are produced by experimental systems that function as "generators of surprise." He continues: "if [scientific activity is] to allow 'new observations' at all, [it] must at first necessarily possess some degree of indefiniteness."

Surprise facilitates new insights and nuanced understandings precisely because it disrupts existing assumptions about psi. In the Ganzfeld experiment, surprise is not merely an outcome of inconclusive results or the milieu in which parapsychologists conduct their research. It is instead something that they can actively build into the design of their Ganzfeld experiment. Experimental subjects play an active role, alongside researchers, in evaluating their own personalities, psi capabilities, and likelihood to identify the target image. While the evaluations of researchers and their subjects may align at times, this "collaborative imagining" can also produce unexpected results when their interpretations come to a head.

As an experimental subject, I am what Sondow considers an "ordinary person," that is, someone who does not self-identify as possessing extrasensory perception.

Surprise acts as a generative force, blurring the boundaries between researcher and subject as both Sondow and I attempt to understand what it means to have a psi experience. Seen in this light, surprise can be seen as creating a space to negotiate and potentially reinterpret the nature of psi itself. It is in these moments that Sondow and other parapsychologists can establish new psi-conducive variables, formulate increasingly subtle provisional theories, or address existing methodological problems within the Ganzfeld experiment.

At a glance, the Ganzfeld experiment shares similar features to other "normal" scientific experiments: controlled designs, double-blind conditions, hypothesis testing. "Academic parapsychologists," as anthropologist David Hess notes, "generally have graduate training and they view parapsychology as a scientific discipline, albeit one that is not generally accepted by the broader scientific community." The most striking difference between parapsychologists and scientists, then, can be seen as their objects of inquiry. Parapsychologists investigate psi, a phenomena that is unpredictable at best and at worst, non-existent.

Through my own studies of paranormal researchers in the United States, I examine how my interlocutors engage with science alongside other knowledge traditions (e.g., religion, occult) as coexisting and complementary resources in order to grapple with the uncertainties of paranormal phenomena. I write about the Ganzfeld experiment because it can tell us a great deal about how scientific novelties occur through gradients of knowing that connect mind and body, reason and emotion, and feeling and cognition. In particular, it can tell us about how scientific procedures, like double-blind tests, are supported by hunches, intuition, and personal transformations. In writing about the Ganzfeld experiment, I believe the most interesting questions do not reside in whether or not it falls within "science" or "pseudoscience." Rather, these labels foreclose more interesting questions concerning their intersecting histories and how they are culturally constructed, coopted by different people with different agendas, and constantly redefined as new ways of thinking about skepticism emerge. And more importantly, they foreclose the possibility that parapsychologists at the margins of science can have anything interesting to say to the broader scientific community, or that the boundaries of what counts as "science" may change in the future. ■

Eliza Cohen

The Biological Facts Committee

One cold morning in February of 1929, five curators at the American Museum of Natural History in New York received a brief memo upon their desks. "Gentlemen," the note reads, "With the approval of the president, I hereby appoint you members of an Advisory Committee on the Biological Aspects of Museum Groups…The purpose of appointing this Committee is to see what can be done to present more biological facts in the groups as now proposed for these halls." The Committee convened several times over the course of the next two years to review the existing and future museum displays. But unlike other formal groups of Museum staff, it is absent from the public reports of the Museum. And two years later, just as suddenly as it was established, it disappears from the archives. It's unclear whether, over the course of their short project, the Committee had any impact at all. But the volatility of their correspondence—indeed, even the existence of the Committee in the first place—provides insight into the Museum's tumultuous decision-making process about how to display biology.

The Committee's directive was deceptively simple: try to re-imagine, from the concept all the way to the display, what kinds of biological information could be conveyed to a general audience. As a vanguard in both research and public communication of science, the American Museum of Natural History played a major role in defining the field of biology. But by 1929, that definition was changing. As the curators attempted to tell increasingly complex stories— including ecological, evolutionary, and behavioral dynamics—they worried that the public would not grasp the information. Simultaneously, as the Museum expanded in size and scope—adding departments, staff, and exhibit space—the central management struggled to coordinate between departments. The Museum's anxiety facing these twin challenges of scale caused them to assemble a formal committee to assess and control the visual storytelling practices in the exhibits. But what was a "biological fact"? Why did the Museum distrust their public? And why was a committee formed when the Museum already relied on the extensive expertise and experience of an entire staff of curators?

In the early Committee proceedings, William Gregory, curator of Comparative Anatomy, pointed out that the curators used the word "biology" in two starkly different ways— sometimes to refer to the physiology and life history of an organism, and sometimes to refer to ecological relationships. In the 1920s, the field of biology as a whole was in a period of redefinition. Biologists wrestled to reconcile 19th century Darwinian evolutionary theory with newer concepts of ecology. Secondary schools had only begun to teach a unified "biology" two decades earlier, condensed from the disparate fields of botany, zoology, and physiology. The Committee formed just four years after the Museum curators concluded their heavy involvement in the Scopes Trial, the first legal challenge to laws prohibiting public schools from teaching evolution. These dramatic national changes forced scientists and laypeople alike to confront the stories told about biological sciences.

As the biological sciences themselves changed, the Museum began trying to tell stories that incorporated more "biological aspects."

Eliza Cohen is from Cambridge, Massachusetts. She is currently an undergraduate studying Science and Society at Brown University. She will give you four nickels for a pair-adigms.

AMERICAN MUSEUM OF NATURAL HISTORY LIBRARY, IMAGES 310841 & 314207

The Committee formed in the midst of a gradual transition into more dynamic, engaging displays. In the 19th century, displays were usually long glass cases of individual fossils and taxidermied specimens mounted and meticulously labeled. But starting in the early 20th century, the Museum began to use more "group displays" which showed specimens gathered together and interacting in order to showcase ecological relationships as well as physiology. In their book *Life on Display* Karen Radar and Victoria Cain have pointed out that group displays were initially viewed as less prestigious, associated more with fairs and other popular entertainment than the austerity of the museum space.

As the curators incorporated group displays, the displays acquired more esteem while allowing the Museum to attract a wider audience with more sensational visual storytelling. In Gregory's mind, the "function" of the museum was to make science "attractive to intelligent people." As he wrote in 1929, "It is not necessary to make science repugnant in order to teach basic principles. Let us rise above dry technicalities and state broad and well established scientific results in simple language and in arresting exhibits." Gregory's "intelligent people" ranged from working-class museum visitors to elite donors.

In addition to attracting a more popular audience, the group displays appealed to the elite as a way to glorify their hunting expeditions by publicly exhibiting their spoils. But not all curators embraced this tenuous relationship. Harold Anthony, curator of Mammology, resented the Committee's description of mammals as "trophy groups," and retorted that Mammology had no display "showing a dead animal with a rifle couched against the beast." He pointed instead to the Department of Marine Life, arguing that their display portrayed a sail fish "vainly endeavoring to shake a hook that is impaled in its jaw. Could any group be more thoroughly a trophy group of a sportsman's display?" The Committee was "diplomatically limited"—they had to placate wealthy patrons who expected the "beauty and grandeur of a large hall with spectacular habitat groups." As Donna Haraway has argued in her essay "Teddy Bear Patriarchy," this often led to a valorization of trophy hunting. In all these debates, the curators balanced their hope "to stimulate biological spirit" with a more sober presentation of biological facts.

As they designed the new group displays the curators were forced to evaluate what they felt the public was capable of understanding in the first place. This was cause for some debate: the complexity of an ecosystem posed an obstacle to clear, simple stories. The curator of invertebrate zoology argued that the exhibits were only capable of showing how specimens interacted in their habitat; other aspects of the natural history, for example, family structure, "might only be brought out in the labels." James Clark, a member of the Committee and head of the Preparation Department, saw display and information as a trade-off, proposing a compromise of "50% biology and 50% picture." Gregory "doubted whether the public saw even the ecological relationships in the mammal and bird groups." The curators were unsure of their power to communicate, and the Committee hoped to direct this process of experimentation.

But some of the curators were incensed— not at the Committee's ideas, but rather at their implication that the curators were not already doing what the Committee was proposing. The curators' resistance to the Committee tells us as much about the process of scientific storytelling as the Committee's suggestions themselves. Roy Miner, curator of Marine Life, asked why the groups are being criticized at all. Harold Anthony agreed with Miner about the role of the Committee, despite his earlier finger-pointing to Marine Life as the real culprit in creating trophy groups. Anthony wrote that the first meeting was full of "destructive criticism based on sweeping generalities." He argued that the curators were already incorporating "many facts of life history, geographical distribution, seasonal variation, etc." The director had to assure Anthony that the suggestions of the Committee did not reflect on his ability as a curator. But Anthony, perhaps worried about his job, wanted to ensure that the process for storytelling was not centralized in any hierarchical bureaucracy. He pushed back against the Committee's hope to develop a generalized policy, and advocated instead for curatorial discretion in determining "the solution of which of these data shall be displayed."

The volatility of the reactions tell a story about the centralization of knowledge production—the Museum used the Committee as a desperate attempt to control and standardize a process and a story of increasing complexity. The Museum functioned as a public institution, and used the same techniques of bureaucratization that corporations were using

AMERICAN MUSEUM OF NATURAL HISTORY LIBRARY, IMAGE 315125

at the time in a desperate "search for order," in the words of Robert Wiebe. As institutions became larger and took on more different functions, bureaucracy—including the technique of forming committees—functioned as what Alfred Chandler Jr. has called a "visible hand" to coordinate the various functions of the institution.

It's impossible to know why exactly the Committee disbanded and disappeared from the archive. Ultimately, it's unclear whether they had any real impact on the museum displays. The Committee focused on incorporating more ecological dynamics, but the archive shows that curators considered ecological dynamics as a major goal in the decade before the Committee was convened. The Committee attempted to search for order over an increasingly complex story, but they struggled to define the meaning of "biology," let alone "fact." They did not find the language to define any sweeping new policy for museum exhibits. With such a broad mandate as determining "biological facts," they may have been using words so big that they meant nothing at all. ■

Christina Agapakis

Conversations with Evelyn Fox Keller

Evelyn Fox Keller is a theoretical physicist, a mathematical biologist, a feminist philosopher, and a historian of science. Throughout her career she has pushed the boundaries of science, confidently crossing the borders that separate disciplines and breaking down the barriers keeping women out of the highest reaches of scientific achievement. Her work has been hugely influential to me, and I was thrilled to have the opportunity to discuss her career with her over the course of several conversations in her Cambridge, Massachusetts home. The following is a heavily edited transcript of our exchange, woven through with excerpts of Keller's writings from different stages of her prolific career.

Even a curriculum vitae is a kind of autobiography. Rudimentary and transparent though it is, it may reveal deeply personal traits. Certainly my own does; it makes abundantly clear that I have something of a problem with borders: in my peculiar psychic and intellectual economy borders are meant for crossing. More, they constitute irresistible lures. I seek them out—not to test their limits but to worry them, as a dog does a bone. Even as a working scientist, I found it hard to stay put, to keep from straying back and forth—in those days between biology and physics, between theory and experiment. And once I strayed beyond the borders of research science, shifted from doing science to writing about it, the problem only grew worse, for now I had many more boundaries to worry."

-Preface to Refiguring Life: Metaphors of Twentieth-century Biology (1995)

Border Crossings

Mostly when I think about border crossings I think about my extra-scientific career. But it is true that within the sciences I crossed many borders. When we talk about my career—it's a thicket. It was a thicket from the get-go.

First I should say, we came from a rather poor, working class family. My parents were both immigrants. I had a brother and sister, both older than me. My sister went off to college when she was not quite sixteen, so when I was twelve and a half. When she came home after her first semester she told me about the unconscious—I thought that was the best idea I had ever heard! I decided then and there that I wanted to be a psychoanalyst. At the same time, my brother thought I was really smart and therefore should become a scientist like him. He was constantly trying to interest me in science, giving me all these books to read. But I had no interest in science, I wanted to be a psychoanalyst! I loved to read, I read all these novels, but I didn't read the books he gave me.

Eventually I went to Queens College and I declared my major as psychology. But I took calculus, I was going to study math no matter what because I liked it, but it had nothing to do with my career. The calculus class met in this huge room with maybe 100 students, and was taught by Banesh Hoffman, who had worked with Einstein.

Here he is greeting his new class, and as he greets every new class, he starts out with a series of trick questions. I'm a New York City kid so I raise my hand and I know all the answers. So he calls me up after class. His first question is "are you related to Maurice Fox?" Can you imagine, this man is teaching four courses a semester, processing all these kids through calculus and he remembers my brother from 11 years ago?

Anyway, I said "Yes."

CDC IMAGE LIBRARY

He says, "What are you majoring in?"

"Psychology."

"Why not mathematics?"

"I don't want to be an accountant."

He says, "What about physics?"

And I said, "What's that?" That was the end of the first conversation.

That same semester I was taking freshman composition and doing terribly. I couldn't get better than a C+. Finally I decided that for one assignment I should do a book review on one of the books my brother had given me. It turns out they were really interesting! One was about relativity and one was about quantum mechanics. I got A's on the reports.

After that I remember going to a party with all these literature majors and someone asked "What are you majoring in?" and I said, "Maybe I'll major in physics, it's neat, and it gets me A's."

He says, "You can't do that, you're a girl!"

I said, "Oh yeah??" That was all I needed. I mean what the hell, I'll major in physics and *then* I'll be a psychoanalyst. That was my plan.

Christina Agapakis is a biologist and writer. She is a founding editor of Method Quarterly.

The Anomaly of a Woman

After I transferred to Brandeis I took one of my first physics courses with Sam Schweber. He came into the room, he turned his back and he started writing on the board. To me it looked like it was in Arabic, some sort of script, some signs and language I didn't know! I dutifully copied them down not knowing, not having clues to what they mean, and spending the weekends taking books home from the library, trying to find what those squiggles meant. It was like a hermeneutic exercise. I was totally engaged. By that summer I had figured out quite a lot.

The second year I'm at Brandeis I'm taking a course in theoretical physics. The third year is my senior year, I do a senior thesis on Feynman's Lagrangian formulation of quantum mechanics—I went fast! And I still had no idea what physics was about. I had no idea—it was to me a formalism. It was like studying a very opaque obscure religious text, and I loved it! What I got from that was the idea that you could deduce, just by thinking, you could deduce the way the world is. I could think about questions like, *Why is space three dimensional?* There were kinds of answers you could come up with, just with thinking in theoretical formal terms.

But it was all very other-worldly; it had nothing to do with the concrete world in which we live. I was totally intoxicated by these ideas, thinking worlds into being, out of pure thought. By my senior year, I thought, *Well maybe I should study physics, really study physics.* Because it was so interesting, I had been so blown away by this idea of thinking the world into being. I decided to go to graduate school and Sam told me, "You have to go to Harvard."

It was really a mistake—a bad bad mistake. Have you read my essay on being a graduate student at Harvard? I won't even talk about it, it was a nightmare. Nobody would talk to me. And I didn't know what to do.

The story of my graduate school experience is a difficult one to tell. It is difficult in part because it is a story of behavior so crude and so extreme as to seem implausible.

Moreover, it is difficult to tell because it is painful. In the past, the telling of this story always left me so badly shaken, feeling so exposed, that I became reluctant to tell it. Many years have passed, and I might well bury those painful recollections. I do not because they represent a piece of reality—an ongoing reality that affects others, particularly women. Even though my experiences may have been unique—no one else will share exactly these experiences—the motives underlying the behavior I am going to describe are, I believe, much more prevalent than one might think, and detectable in fact in behavior much less extreme…

Sometimes I was queried about my peculiar ambition to be a theoretical physicist—didn't I know that no woman at Harvard had ever so succeeded (at least not in becoming a pure theoretical physicist)? When would I too despair, fail, or go elsewhere (the equivalent of failing)?

-"The Anomaly of a Woman in Physics" (1977)

The summer after my first year my brother and his family were going to Cold Spring Harbor and there was a spare bed in the baby's room. I could come and spend the summer in Cold Spring Harbor and figure out what I was going to do. I arranged it so that I could keep getting my graduate student stipend: I did a reading course on interpretations of quantum mechanics. I packed a suitcase full of Freud, and went to Cold Spring Harbor.

So there I had the quantum mechanics, the psychoanalysis in my suitcase, and there I was at the birth of molecular biology. There were all these young guys who were just going to remake the world. It was so interesting; I got seduced and brought into the lab. I was working with Frank Stahl and Max Delbrück—everyone was trying to lure me into molecular biology. That was the time! I was young, I was attractive, I was smart, and they were all looking for new recruits. I was hot.

In Frank's lab I discovered a phenomenon. It was a very simple idea, but I recognized its power and that was really something. We didn't know the

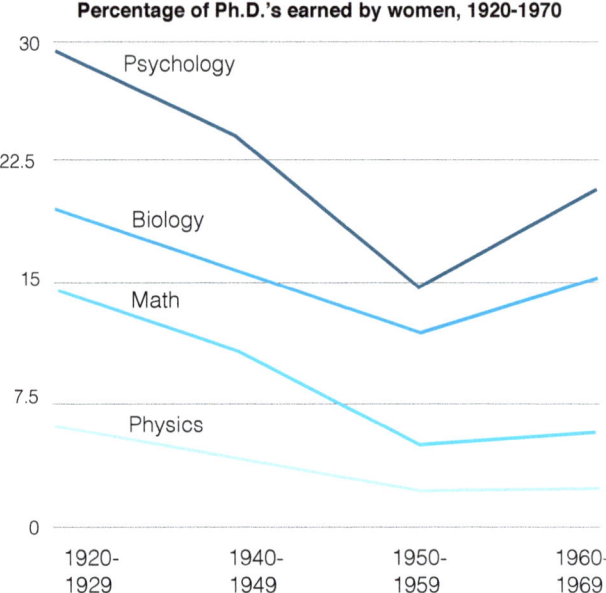

Percentage of Ph.D.'s earned by women, 1920-1970

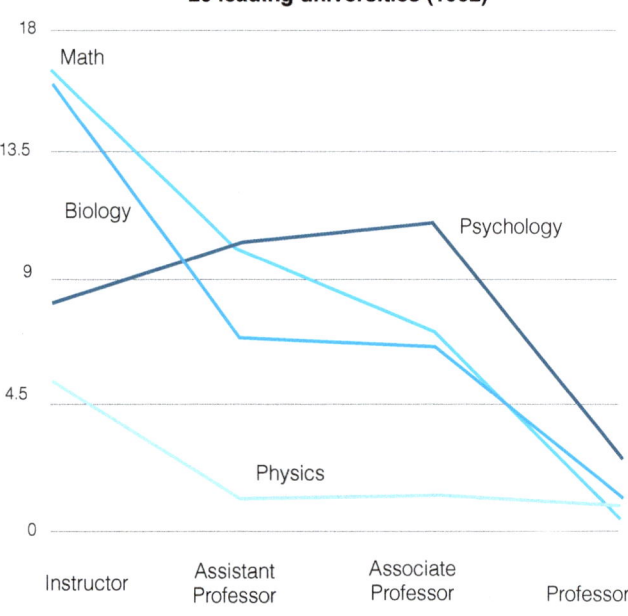

Percentage representation of women, by rank, in 20 leading universities (1962)

Adapted from "Women in science: An analysis of a social problem," by Evelyn Fox Keller, Harvard Magazine (1974).

genetic code yet, and I recognized a way that we could use a chemical variant of DNA to answer the question of whether one or both strands were coding for proteins. *What about the complementary strand, is that coding for the same protein, or is it not read?* I could do this for a thesis and I could do it in three months—it was easy! *Biology was so easy!* It didn't turn out that way of course, but that's what got me into it.

The Question of Gender

I remember a time when I was talking with one of the professors in the physics department, before I switched to molecular biology. He had been friendly, and I remember getting a ride home with him from a seminar at MIT, and he asked "How's it going?" and I started to tell him about how no one would talk to me—everything that eventually became "The Anomaly of a Woman in Physics." I tell him and he looks at me as if I'd just committed a terrible faux pas. I opened the car door and I walked out.

I didn't talk about it after that—at least not until 1970, when I was invited to give a series of talks about mathematical biology at the University of Maryland. I decided that I couldn't in good conscience give five lectures on my work in mathematical biology and never mention the fact of being a woman. So I write up four talks on mathematical biology and for last talk, I talk about women in science.

I didn't talk about it after that.

In that talk, I didn't just talk about my own experiences. In the few years before that, I had started thinking about why there are so few women in science overall and I had started collecting data. I started out this lecture with an equation—a birth and death equation—and I showed the statistics I had gathered about how the percentage of women drops off dramatically throughout the academic ladder and I asked, *What is killing them off?*

The rest of the lecture was a review of all the barriers that operate for women in science. In the end I concluded that probably the most important factor is the widespread belief in the intrinsic masculinity of science. What is that belief doing in science? And what effects does it have? First, on women who try to be scientists, and second, on the science. We're not talking about the influence of *sex* on the science, we're talking about the *belief.* We're talking about gender as an ideology.

I wrote up that talk and published it in *Harvard Magazine,* and shortly after that I published "The Anomaly of a Woman in Physics." I remember after that was published getting on an elevator packed with men, male scientists, and one female colleague. She said "oh Evelyn, how does it feel to go naked in public?"

At that time it was a taboo subject. Actually, if you look at the history of women in science, women in science in the first part of the century were doing relatively well. The real dive came with World War II and it coincided with the strategy of removing their first names. Because they didn't want attention called to them as being women—because they wanted to be just neutral—they took their names off. There were multiple times when I tried to talk to people who had written or spoken about my work but they would ignore me, not realizing that *I* was EF Keller and that I was a woman!

> *We are speaking of course of the problem that inheres in the fact that the universal standard is after all not neutral—of what happens to our strategy, and our thinking about gender and science when we begin to notice that the universal man is, in fact, male. The first thing this recognition enables us to do is to begin to make sense of the failure of the promise of equity. To be included in the big "one" means not to be equally represented, but to be unrepresented. It is to be swallowed up whole, negated in the quest for assimilation—as it were, a hole in "one".*
>
> *-"How Gender Matters, or, Why It's so hard for Us to Count Past Two", in* Perspectives on Gender and Science, *ed. Jan Harding (1986).*

Stereotypes are Never Idle

When people would ask me what I was working on, I'd say, "Gender and science."

And they'd say, "What is it you've learned about women in science?"

"I haven't learned anything about *women* in science—if anything I'm writing about *men* in science and their beliefs!"

But nobody got it, and everyone kept insisting I was writing about women being genetically different, or women doing a different kind of science. No! I was writing about doing science in a gendered world. This was very frustrating, because the collapse of my argument from *gender* to *women* made my argument very different—politically as well as conceptually. It made it seem like I was making an argument for women not belonging in science, because if science is masculine then women don't belong there. I was very sensitive to that.

> *(Because confusion is so easy here, let me make it absolutely clear that this is not an argument for duality—for a female science either to replace or supplement a male science. That is, I do not believe that science is written in our chromosomes—neither in the X nor the Y chromosome.)…*
>
> *The relationship between gender and science is a pressing issue not simply because women have been historically excluded from science, but because of the deep interpenetration between our cultural construction of gender, and our naming of science. The same cultural tradition that names rational, objective, and transcendent as male, and irrational, subjective, and immanent as female, also, and simultaneously, names the scientific mind as male, and material nature as female…Modern science is constituted around a set of exclusionary oppositions, in which that which is named feminine is excluded, and that which is excluded—be it feeling, subjectivity, or nature—is named female. Actual human beings are of course never fully bound by stereotypes, and some men and some women—and some scientists—will always go beyond them. But at the same time, stereotypes are never idle.*
>
> *-"How Gender Matters, or, Why It's so hard for Us to Count Past Two"*

Talking About Language

Much later, in the mid 80's, I was invited to give a talk on gender and science at a colloquium that was co-sponsored by the Departments of Women's Studies and the Physics Department. I walk in, and the room is packed. Turns out, it's the day they award the undergraduate prize in physics, so the whole physics department was there. Big room, and it was completely filled—all men—except in the front seats, on the right, there were four or five women from Women's Studies.

So I gulped and realized this was going to be a tough challenge! I gave my talk, and I talked about the idea of gender in science. At the end one of the physicists said, "Yeah, but you're not talking about physics, you're talking about language!"

That was interesting. I agreed; I was talking about how the language of science was shaped by our assumptions about gender, our model of science on gender, and the use of gendered metaphors in science. My assumption was that our metaphors and the way we talk about science affected the science we did. I was operating on the assumption that language affects the way we think and what we do, and this was the first time I realized that wasn't obvious.

That marked a turning point from my focus on *gender* and science to *language* and science. I wrote on metaphors in evolutionary biology related to competition and reproduction, and I also wrote about developmental biology.

Delayed Effects

Developmental biology claims the interest of feminist historians of science on three different grounds: it is a field in which women have historically been relatively numerous, and in which a number of women today are leaders; in large part because of its intimate association with reproduction, traces of implicit and explicit gender coding can be found in the historical structuring of the field and hence can be used to illustrate more general arguments about the symbolic work of gender in the natural sciences; and the fundamental problem of developmental biology resists resolution in terms of "master molecules" and seems to require, instead, conceptual models of just the kind that contemporary feminists have shown partiality to—that is, models of complex interactivity.

-"Developmental Biology as a Feminist Cause" (1997)

The first thing that interested me about developmental biology was that it just exploded—it burst on the scene in the early '90s, and it made me conscious of the fact that the questions about development had been buried by the onslaught of molecular biology. The subject had originally been called embryology, but the departments of embryology through the '40s '50s and '60s were dying like flies. Molecular biology had such extraordinary successes, and problems with embryology were so difficult and could claim so few dramatic successes, that the limelight just shifted away from it.

One of the main triggers in the re-emergence of developmental biology in the late '80s and early '90s was the work of Christian Nüslein-Volhard. She showed how it was possible to study the embryological development of the fruit fly.

The Drosophila embryo is a boundary object par excellence, residing in the interstices of two major disciplines, genetics and embryology. Its singular value both for the study of development per se and for its integration with the study of genetics is now widely recognized. But it was not always so. The history of Drosophila embryology exhibits not only the strengths of boundary objects in their capacity to facilitate interaction and collaboration, but also the equally obvious weaknesses that can result from their marginality.

-"Drosophila Embryos as Transitional Objects" (1996)

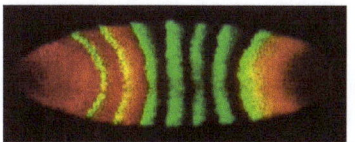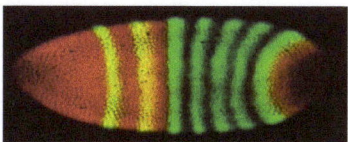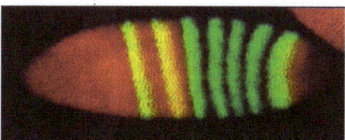

The expression of genes in Drosophila embryos is patterned in stripes across the length of the embryo according to gradients established in the egg. Image adapted from Bergmann et al. PLOS Biology, 2007.

With those tools, she could understand how genes were involved in development. Before her work, the assumption was that history of the organism begins with fertilization and ends with maturity: at fertilization, the sperm enters the egg and "activates" the genes. The sperm puts the genes into action—I'm trying to spell this out explicitly—and it triggers a chain of reactions that results in the adult. What Nüsslein-Volhard focused on with these maternal mutations were effects that were set in motion prior to fertilization: the topological map that was established in the egg prior to fertilization.

She established what's happening in the egg, which is purely maternal, and that manifests itself in the development of the stripes in the adult fly. Speaking of language, earlier they had called these "delayed effects." They put them not *here*, in the egg, but *there*, emerging from gene action, from fertilization.

It was a different frame that led to a different conceptualization—not of gene action but a much more dynamic, interactive view. This was a worldview of development that was more sympathetic with feminist interests and feminist philosophy. But is it that because they were women? I didn't think so. Is it because they were reproductive? I didn't think so. It was outside the box, it was open, and it allowed one to look at other dimensions of the problem. That was welcome to feminists, for all kinds of reasons, but it would have been welcome to lots of other people too, who had nothing to do with feminism.

The Humility of Complexity

I think today is the most exciting time in biology, certainly in my lifetime. All kinds of assumptions that we made because they just seemed obvious are being overturned, in a way that's very, very exciting to me. A lot of it is discovering—which has been the history since the early successes of molecular biology—more and more complexity in the system that is often beyond our capacity to make sense of. So it gives us kind of humility. It also raises some questions about what would a theory of biology look like, I mean what makes us think that we are *capable* of a theory of biology. Our brains are parts of biological evolution. I don't see any reason to assume that our brains are capable theorizing the levels of complexity that biological evolution has given rise to—in our mental capacities for example.

In the early '90s, when I had gotten interested in developmental biology and I shifted my research focus, I was just so blown away by how wonderfully complex biology had become. Compared to the way we talked about genes, it was a joke. After a particularly good series of lectures at Caltech I remember asking one of the biology professors, "Why did you leave out all the problems?"

He said, "Well it would be too complicated for a lay audience—they wouldn't understand."

So I resolved to talk about genetics as it was currently being addressed, in the ways that I saw, in all its complexity. I said, "I can make it understandable."

It is a rare and wonderful moment when success teaches us humility, and this, I argue, is precisely the moment at which we find ourselves at the end of the twentieth century. Indeed, of all the benefits that genomics has bequeathed to us, this humility may ultimately prove to have been its greatest contribution. For almost fifty years, we lulled ourselves into believing that, in discovering the molecular basis of genetic information, we had found the "secret of life"; we were confident that if we could only decode the message in DNA's sequence of nucleotides, we would understand the "program" that makes an organism what it is. And we marveled at how simple the answer seemed to be. But now, in the call for a functional genomics, we can read at least a tacit acknowledgment of how large the gap between genetic "information" and biological meaning really is.

-The Century of the Gene, excerpted in the New York Times *(2000)*

■

Patrick Boyle

Going Viral

A virus races through densely populated China, and the numbers are grim: mortality rates north of 80%, and millions dead in the first year of the outbreak. Three years later, the epidemic reaches the United States, spreading to 16 states in three months. DNA sequencing proves that the American version of the virus originated in China. It seems that since it made it into the U.S., the virus was rapidly disseminated by contaminated transportation infrastructure. With no known cure for the virus, the focus shifts to isolating individual cases while the epidemic runs its course.

USDA

This devastating epidemic is very real and ongoing. First identified in the United Kingdom in the 1970s, today's Porcine Epidemic Diarrhea Virus (PEDV) outbreak began in China in 2010. We are still learning the extent to which our globalized infrastructure helped this epidemic spread. In the meantime, PEDV has driven US pork prices up more than 13 percent in the first year of the epidemic. To offset the loss of pigs killed by PEDV, growers are increasingly relying on imports from Europe. While this solution is only temporary, it has helped lessen the impact of PEDV on retail pork prices—and so has raised little concern from consumers, most of whom don't know we are in the midst of an epidemic.

Fortunately, PEDV does not appear to infect humans, but it is a close cousin to other viruses that have made the jump across species boundaries including Severe Acute Respiratory Syndrome (SARS) and Middle East Respiratory Syndrome (MERS). Each infected organism (animal or human) offers a virus another billion chances to evolve and change. The H5N1 and H7N9 "bird" flus as well as H1N1 "swine" flu first infected workers caring for birds and pigs. These flu strains have an unusually high mortality rate, but appear to not spread effectively between people; in these cases, the animals act as "reservoirs" for the disease. In public health circles, this realization has led to the "One Health" concept, which acknowledges that managing and treating disease in animals is as important as preventing disease in humans.

Early understanding of the link between animal and human health occurred hundreds of years ago, when the decision to move livestock outside of people's homes revolutionized public health. Today, however, modern infrastructure has counterintuitively strengthened the link between animal health and human health: more people means more domestic livestock, and global trade networks offer ample opportunities for the spread of disease. Furthermore, today's Ebola epidemic has vividly illustrated that the spread of disease through urban centers has global consequences. Ebola outbreaks, which are thought to originate from bats, may be more difficult to limit now due to increasing urbanization. Both globalization and climate change may be contributing to the spread of insect-borne diseases such as West Nile virus and Chikungunya virus, which were once thought to be restricted to tropical regions.

Animal populations also play a major role what many consider to be one of the greatest health threats in the 21st century—the growing ineffectiveness of antibiotics, one of the 20th century's greatest health advancements. Clinical overuse of antibiotics has played a well-known role in the rise of antibiotic resistance. Somewhat lesser known is that most antibiotics manufactured every year end up in animals, not humans. The FDA estimates that about 8.3 million kilograms of "medically important" antibiotics (meaning drugs similar to or identical to antibiotics used for humans) were used for use in food-producing animals per year. Over the past half-century, antibiotics have been added to the feed of healthy farm animals. For reasons that are not entirely clear, this helps animals grow more quickly. Efforts to curb this practice, particularly in the European Union, have faced many challenges: antibiotics are cheap, and the conditions in modern farms often lead to animals getting sicker when antibiotics are removed. In the meantime, the effectiveness of antibiotics in the clinic is undergoing a steady decline.

Keeping up with these emerging health threats may require radically new approaches to drug and vaccine development. Complete physical containment of biological threats that have reached animal populations is impossible; improving farm practices and human behavior doesn't prevent the spread and evolution of diseases in wild animals. As an alternative to physical containment, researchers are exploring the potential for biological containment: can we engineer biological systems that limit the spread of disease in animal reservoirs? If so, we could fight diseases when they occur in animals, improving the health of livestock and limiting animal-to-human disease transmission.

■ ■ ■

The field of synthetic biology [full disclosure: my area of research] may offer new approaches to managing diseases that spread through animal populations. Synthetic biologists are biological engineers. The world around us provides countless examples of how biology is a unique and powerful engineering substrate. It's hard to imagine a micron scale robot that derives all of its power from the local environment and can make copies

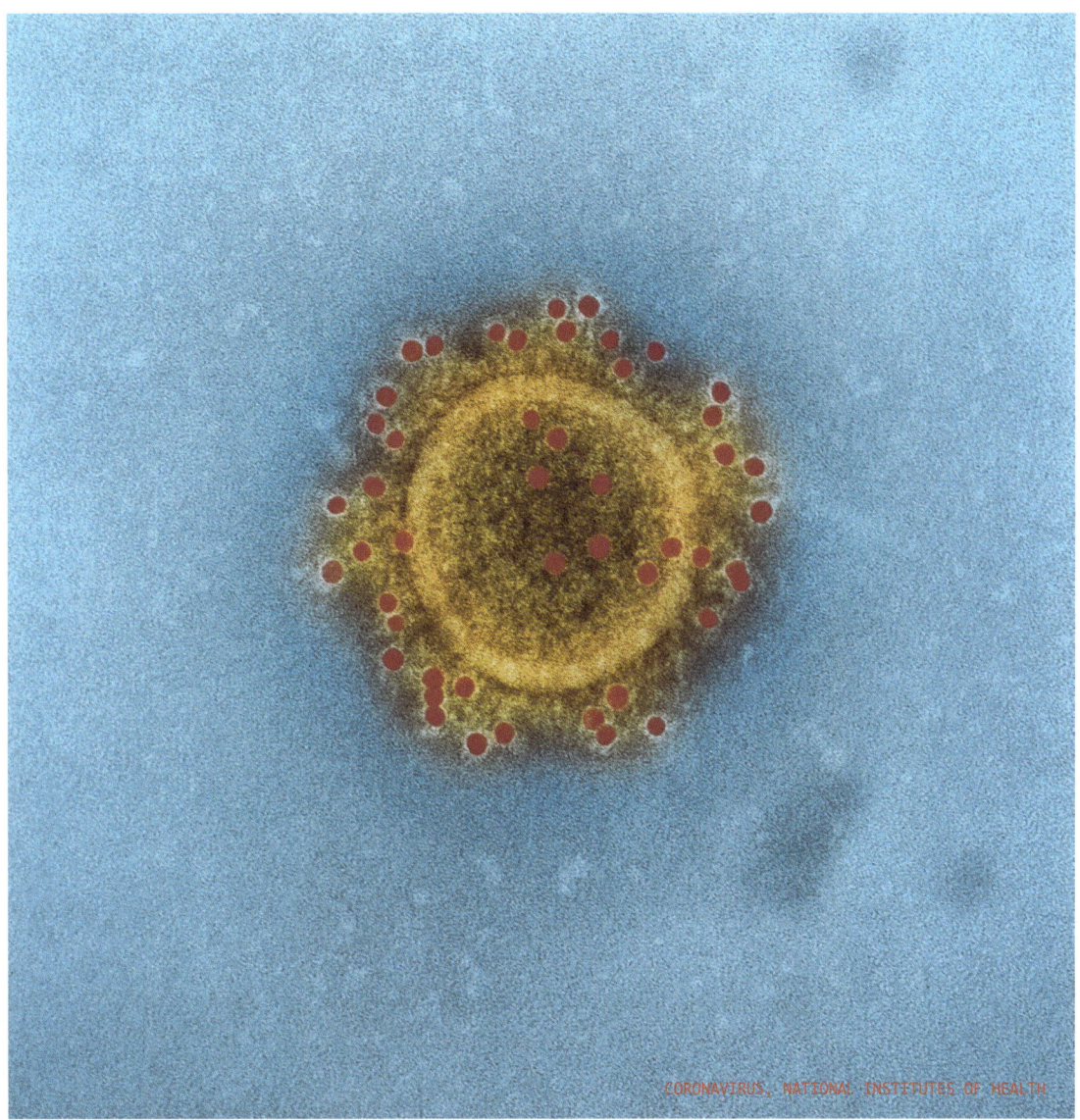

CORONAVIRUS, NATIONAL INSTITUTES OF HEALTH

of itself every 30 minutes, and yet trillions of bacteria in your gut do just that. Synthetic biologists hope to tap into the power of biological systems to make new medicines, vaccines, chemicals, foods, and all the other products that biology is good at making.

Simply developing new tools for engineering biology is not enough to enable new approaches to fighting disease. These tools literally need a place to live. Synthetic biologists are engineering organisms that can thrive in specific niches, including the gut microbiome. In people and animals, one of the richest niches for microbes is the gut. The trillions of microbes living in your gut aid in digestion and help protect us from the few microbes that can make us sick. How might engineered bacteria in the gut help limit the spread of infections?

Gastrointestinal infections like PEDV are often transmitted through the "fecal-oral" route: food or water contaminated by the feces of a sick individual. Bacterial hemorrhagic diarrhea, like that caused by *E. coli* O157:H7, is often acquired through meat tainted in farms and slaughterhouses. Engineered microbes could help us protect ourselves from these infections before they make us sick. Better yet, livestock "vaccinated" with an engineered microbiome of their own could block these bad microbes at the source—farm animal feces.

Engineered microbiome research is in the early stages. The first step towards microbiome treatments, fecal transplants, have had significant success in early clinical trials—good microbes from a healthy donor can "boot up" a functioning gut microbiome to crowd out the bad microbes that were causing infection. The biggest role for fecal transplants today is to help patients recover from *Clostridium dificile* infections. *C. dificile* can take over the gut when other competing microbes have been wiped out by antibiotics; fecal transplants can supply enough good microbes to take that territory back. Of course, it would be better to avoid gut infections in the first place. Synthetic biology researchers have also been able to engineer normal gut bacteria to sense changes in the gut environment in mice. Advanced versions of these engineered bacteria could someday stand guard in your gut and administer therapeutics if an infection or other problem is detected.

Synthetic biologists are also exploring how engineered microbes may be able to protect against antibiotic resistance. Bacterial infections are not always caused by "bad" bacteria; many infections (including *E. coli* O157:H7) are caused by "good" gut bacteria that have acquired bad genes. Genes conferring antibiotic resistance in particular can spread rapidly through populations, passing between different individuals in a microbial population. Early experiments with engineered proteins that can target and cut specific DNA sequences have shown that it may be possible to target and eliminate these antibiotic resistance genes from a population, stopping the spread in its tracks.

■■■

Microbiome treatments may transform how we treat disease in people and farm animals, but there will always be diseases that incubate and evolve in wild animals. In the case of insect-borne diseases like malaria, newly published research has shown that a naturally occurring species of

bacteria living in the gut of mosquitoes might be able to prevent the malaria parasite from infecting the mosquitoes. Feeding mosquitoes probiotic nectar with this species significantly reduces their susceptibility to the infection, making them less likely to spread malaria to humans.

Another radical (yet increasingly feasible) approach to managing insect-borne disease is to engineer the mosquito populations themselves through genetic elements known as "gene drives." Like viruses, gene drives are selfish genetic elements. While normally offspring receive one copy of each gene from the father, and a second copy of each gene from the mother, with gene drives, the gene in question can duplicate itself and erase the copy of the gene from the other parent—allowing a new gene to spread through a population much more rapidly.

In the case of malaria, mosquitoes engineered with gene drives that cause sterility in offspring could mate with wild mosquitoes. As the sterility genes spread, the mosquito population would decline, and eventually be eliminated from the ecosystem. Others have proposed designing gene drives that protect the mosquito from being infected with the *Plasmodium* parasite that actually causes malaria, leaving mosquito populations intact.

Gene drives have not been tested in the wild yet; however, other tools for controlling mosquito populations have undergone limited field trials. One technique involves the release of sterile male mosquitoes into wild populations. These males mate with wild females and reduce the number of offspring in the next generation. This has been done for decades via the use of radiation or chemicals to sterilize the males; over the past decade more precise bioengineering tools have led to engineered male mosquitoes with more predictable traits.

The development of this technology is illustrative of the rapid pace at which biotechnology is advancing. The gene editing technologies essential to building gene drives are the same as those used to target antibiotic resistance genes in the microbiome. This technology was first described in early 2013. About a year later, papers extending

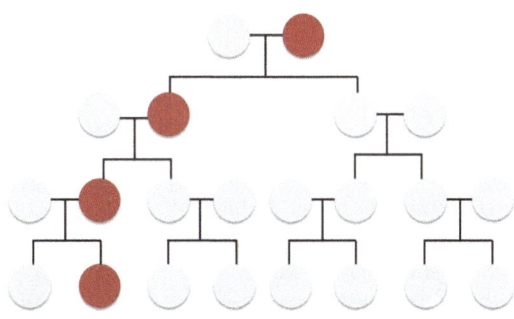

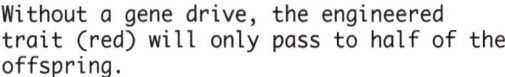

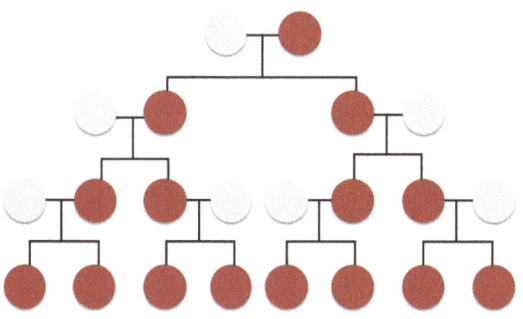

Without a gene drive, the engineered trait (red) will only pass to half of the offspring.

With a gene drive, the engineered trait passes on to 100% of the offspring and rapidly spreads through the population.

this technology to microbes, plants, and animals have been published, including one where researchers in China successfully edited the genomes of monkeys. The cost of reading and writing DNA has become so cheap that new bioengineering advances can spread virally (in the internet sense of the word), with researchers synthesizing new genetic circuits based on the latest advancements around the world. This ease of information sharing and DNA synthesis represents a fundamental change in how we do science, and how quickly new techniques and designs can disseminate among researchers.

■ ■ ■

Caught between the evolving threat of animal-borne disease and the rapid development of biotechnology to meet these threats lies a patchwork of regulations designed for 20th century biology. In the United States, many regulations are based on the control of agricultural pests or diseases and were not originally intended to regulate engineered organisms. For example, if you want to engineer plants, the tools you use (rather than the changes you make to the plant genome) determine who regulates your new plant. Many researchers use a bacteria called *Agrobacterium*, which is good at taking a specific piece of DNA that you give it and transferring it into the plant of your choice. If you want to sell the engineered plant, the USDA's Animal and Plant Health Inspection Service (APHIS) has regulatory jurisdiction because wild *Agrobacterium* is a plant pathogen. If you use a gene gun (literally a gun that shoots DNA into things) to put the same piece of DNA into the same plant, you are exempt from APHIS oversight. This theme is repeated across the regulatory system, with policies generally concerning specific methods rather than the end result. In today's world, where new methods develop in a span of months rather than decades, these regulations quickly become outdated.

But while a gene drive trial in the United States could conceivably fall under the jurisdiction of the FDA, USDA, and EPA, it and other environmental applications of synthetic biology are likely to cross regulatory and international borders. Internationally, field trials with engineered mosquitoes have been performed under geographic isolation (usually on islands) and under the specific laws of the local country. A gene drive-based solution could have a larger geographical range, requiring multinational cooperation for regulation and oversight.

Of course, regulations do not prevent people from engineering biology. As with other technologies, advances in biotechnology are becoming increasingly accessible to the general public in addition to skilled researchers. While you can't exactly build a gene drive in your basement, there is a growing community of hobbyists in the synthetic biology world. To adapt to this leveling playing field, we will need a better understanding of the risks posed by environmental deployment of biotechnology, and cultivate communities that encourage responsible technology development. Fortunately, these issues are being discussed, with efforts to include input from all of the relevant stakeholders. For example, the Synthetic Biology Project, led by the Woodrow Wilson International Center for Scholars, has convened a series of workshops that bring synthetic biologists, environmental scientists, regulators, hobbyists, and concerned NGOs together to discuss new case studies. These case studies are important in focusing the discussion to specific projects that are likely to be field-tested in the near to medium term, helping to separate the real elements of new technology from the hype—both positive and negative—to make informed decisions. These case studies also help the synthetic biologists involved in these projects take input from stakeholders, and adjust the design of their project to better address their concerns. In a series of publicly available reports, the Synthetic Biology Project has outlined research and regulatory priorities relevant to helping us understand and evaluate new synthetic biology technologies.

■ ■ ■

Improving human health in the 21st century will not be driven entirely by biotechnology—culture, politics, and values play a large role in how we manage disease—but we won't be able to simply rely on the drugs and practices that got us here. Given the myriad emerging health threats we face, it will be a challenge to maintain the current health standards of the developed world over the course of this century. Antibiotic resistance can't be addressed by continuing our current widespread use of antibiotics in agriculture; we will likely need a mix of new practices and technology to save the antibiotics we have left. PEDV has shown us that borders and high local standards don't shield us from global epidemics. Improving human health in the future will mean improving global health.

While we face immense challenges, we also have unprecedented access to tools and data that can help us adapt to these new realities. Gene sequencing and data sharing enabled pig growers and researchers to trace PEDV to its source and slow the spread of the epidemic. Synthetic biology could allow us to vaccinate pig populations against future PEDV outbreaks, either via genome editing or a microbiome upgrade. Careful and regulated tests of these technologies will also give us a better understanding of how diseases spread and persist, since we will need to adequately quantify the effectiveness of novel interventions. Developing tools that can address human diseases in animal populations would be a breakthrough on par with the discovery of antibiotics: a tectonic shift in how we fight infectious disease. ■

Patrick Boyle is in charge of the Design pipeline at Ginkgo Bioworks, a company that makes and sells engineered organisms

Thalia Field

He Told Animal Stories

Here Am I—Where Are You?

Young students hear the word *imprinting* and imagine the way motherless ducks in a moment of infancy are said to form an irreversible bond to whatever stands in the place of a parent.

Konrad Lorenz receives a Nobel Prize for physiology and medicine in 1973; a white-bearded man in work pants and waders, a row of ducklings strolling behind him.

> Lorenz: *"Without supernatural assistance, our fellow creatures can tell us the most beautiful stories, and that means true stories, because the truth about nature is always far more beautiful even than what our great poets sing of it, and they are the only real magicians that exist." (King Solomon's Ring)*

Picture: Konrad Lorenz on his steps, feeding a baby bird from a dropper; Martina the goose waiting to go up to sleep in 'her bedroom' at the top of his house; a family portrait in progress.

Omissions and slips, folds and puns—people struggle to reconcile how from the smallest tropes evolution can be described.

For example, one summer morning in Vienna, a fly either was or was not on the kitchen ceiling when a hand-reared starling made a grab for it. A long winter of waiting had built up in the starling and it couldn't help itself. Fly or no fly, it was going to eat one.

For Konrad Lorenz, this starling's *vacuum activity* happened in response to nothing, and in happening, told an instinctual story. Lorenz told stories about why people do many things, based on animal analogy.

Analogy:

> 1. *inference that if two or more things agree with one another in some respects they will probably agree in others*

> 2. *resemblance in some particulars between things otherwise unlike; comparison based on such resemblance*

3. correspondence between the members of pairs or sets of linguistic forms that serves as a basis for the creation of another form

Related of course, metaphor: the truth in an erroneous naming.

Aristotle: "Metaphor consists in giving the thing a name that belongs to something else."

But where a metaphor at some level acknowledges *an intuitive perception of the similarity in dissimilars,* an analogy presumes a more correct correspondence. For evolutionary biologists, analogy implies equality in function, proximal cause. Where metaphors overreach, analogy occurs frequently.

Disturbance of Characteristic Behaviors through Domestication

One morning at his father's house on the warming side of winter, Konrad's hand-raised starling flew suddenly to the ceiling near the window. Lorenz was *sure* the bird caught what wasn't anything, flew back to its perch, decisively killed, ate, and swallowed the imaginary meal.

But what if there was in fact a fly?

No, Lorenz insisted there had been *nothing* at the ceiling near the window, and the bird had merely imagined something, acting out an unreleased instinct, in what Konrad forever swore was clear proof of *action-specific potential.*

There had been no fly, Konrad was sure. The bird simply needed there to be the fly, fantasized the fly, the imagined taste, the feeling of it in its beak and throat. *In vacuo—auf Leerlauf—*this starling and the fly-that-wasn't-there made the perfect team.

Lorenz: "Although it had never trapped a fly in its whole life, [it] performed the entire fly-catching sequence without a fly."

A skeptical colleague once asked Lorenz, "Is that something that actually happened or just something you saw?"

In other words, is storytelling your scientific method?

For his Nobel prize acceptance lecture, Lorenz presented the paper: *Analogy as a Source of Knowledge.* He showed military bombers next to birds, torpedoes next to sharks, to show how physical features are analogous.

Is storytelling your scientific method?

And he told about Bernard Hellman his "important childhood friend." Lorenz and Hellman were born in Vienna on the very same day, in the same year. Their families had apartments in Vienna and neighboring homes in Altenberg. The boys attended the same Gymnasium, explored the same ponds and riverbanks. Bernhard was no mere Kumpan, but a friend close enough to type up Konrad's diary about jackdaws and send it to an editor—for what became his first publication. Bernhard was known as the brighter student and biologist, and also as the constant third-wheel with young Konrad and his future wife, Gretl, taking hundreds of excursions around Europe on their 500cc overhead-valve Triumphs.

The Rise and Fall of Man and Animal

It is well known that Konrad Lorenz considered himself the "father" of animal behavior studies (ethology). What can it mean to be the father of something, to lend part of yourself to a new being, to stand ready and responsible? The fathers of the country, the fathers of destiny?

Lorenz wanted to be father and mother, believed in having a father, following fathers, becoming what fathers want, which was to mother, in their fashion. The law of the father, the jungle, must be followed. But to take the place of the mother, to experiment on hatchlings, to make children your project, this was more than most fathers get to do.

What a Pity He Can't Speak! He Understands Every Word!

Konrad Lorenz was reported to be charming enough that after 'the terrible war' his colleagues forgave him his collaboration with the Nazis.

Even Niko Tinbergen—the other 'father' of ethology, the 'nice father'—the suffering, persecuted father—helped Konrad Lorenz regain his "scientific children" in the court of public opinion. But for all Tinbergen's generosity toward him, Lorenz could never bring himself to apologize for, or recant, his Nazi-sympathetic writing. As Tinbergen later wrote to a mutual friend: "at least once say 'sorry brother, we are very sorry about it all and we will never support such a gang again.'"

Analogy: a foreign name; the wrong word in place of the usual.

Why is analogy appealing? To decorate an idea. To impress or persuade. To generate a truth which lies beyond the familiar or bring into view something which never previously existed. Some say an analogy is structured like a joke, in which one recognizes it and then decides whether to go along.

> Lorenz: *"Why should not the comparative ethologist who makes it his business to know animals more thoroughly than anybody else, tell stories about their private lives?"*

Konrad knew why jackdaws or greylag geese did things because he assiduously recorded the ritualized bowing, strutting, egg-rolling, courtship, neck-flexing. *The Year of the Greylag Goose* describes thousands of details about neck and wing position. And by analogy why some people are naturally superior to others.

> Lorenz: *"As Rudyard Kipling's Mowgli thought of himself as a wolf, so Tchock, had he been able to speak, would certainly have called himself a human being."* (Breakdowns in the Instinctive Behavior of Domestic Animals and their Social-Psychological Meanings)

PHOTOGRAPH OF KONRAD LORENZ FROM POW CAMP

Turning his mansion into a home for birds, "Mother Konrad" slept under the eaves listening to the scratching geese and jackdaws making themselves at home on the rugs and rafters. More than anything, he wanted to catalogue "ethograms" of wild animals in "natural cage conditions," even placing his own children in "reverse cages" to protect them. Simply to make his birds' repertoires available for controlled observation was the purpose of hand-raising them. According to Lorenz, the laboratory is too small, the whole world too wide, and the zoo too open to the public. Marble halls painted with cherubs and adorned with Greek statues provided the perfect habitat for his geese.

> Lorenz: *"By keeping a living thing in the scientific sense we understand the attempt to let its whole life cycle be performed before our eyes within the narrower or wider confines of captivity."* (The Year of the Greylag Goose)

When the Nazis came to power, Lorenz was given his first real job, a position in Konigsberg that most saw as political pay back. In 1941, Lorenz was drafted to the army, and his official story was that he worked as an army doctor until being held as a POW by the Russians for four years.

How is a story not the story? What if there was a fly?

In the records, there is a missing period of six months in 1942, during which time Lorenz had no official duties in the German-occupied Polish town of Poznan. However, the published work of a man named Hippius refers to Lorenz participating as a psychological evaluator for a group of 877 children—*mischlings*—of mixed Polish/German heritage. His job required him (with his background in domestic and hybrid wild geese) to discern the psychological make-up of the children, a form of *volkerpsychologishe*—looking for traits of national character—assuming that hybrid people become detached from pure parental values. As the Germans believed in the importance of race separation for animals and people alike, sufficiently "German" *mischlings* could be resettled in the new Germanizing east, and those without patriotic profiles ("instinctual cripples" as Lorenz once called them) were sent to concentration camps.

Lorenz, 1940: "With phenotypic inferiority the refined modes of social behavior are disturbed far earlier and far more seriously than the outward appearance. One can predict with absolute certainty of a crooked-legged, pot-bellied, pale-beaked grey goose, such as is all too easily produced through careless breeding, that its social behavior will be other than normal. With the pure-blooded wild goose the view of the old Greeks that a handsome man can never be bad and an ugly man can never be good is fully valid."

Is biology condemned, through its use of figuration, to be literary, and somehow the reverse is equally true? Does writing reveal biological truths of individuals, worlds, how ontology relates to stories, their biological basis, their evolution? Scientists engage with silent yet definitive symbols: fish who switch sexes when conditions call for it, super-organisms with a socialist agenda, or viruses on computers. Enhancing the world through analogies—or catachresis, or prosopopeia (it feels natural to include ghosts and monsters)—brings new things into being, whether or not they are real.

Reading science as biography or poetry feels both rich and problematic. Biologists mostly work and write quantitatively, to loosen language's messy involvement. Also, writing from the perspective of species rather than individuals avoids psychoanalysis, or the overlapping of language and fantasy. But some biologists now consider living beings merely diverse processes of semiosis. To the mind, the inexhaustible displays of our earth will model anything. Plotting and counting ants as they make their way across a square-foot patch of sand, the students can't remember what they are supposed to be seeing, or doing. Research frustrates language, and vice versa.

Behavioral Analogies to Morality

Lorenz: "Real friendship with wild animals is to me so much a matter of course that it takes special situations to make me realize its uniqueness." (The Companion in the Bird's World: Fellow Members of the Species as Releasers of Social Behavior)

After the war, Lorenz published his most popular book, *King Solomon's Ring*, a stark effort to rid his work of propaganda and return to himself the reputation of the animal-loving scientist. Stories from before and during the war, about the losses of jackdaws, dogs, and geese, fill the pages. About friendship with dogs, Lorenz ponders, "…we may well ask ourselves whether we do right to hang our hearts on a creature which will be overtaken by senility and death before a human being, born on exactly the same day, has even passed his childhood."

Only one passage implies a wider context: "A past master of this art [aquarium planning] was my tragically deceased friend Bernhard Hellman who was able to copy, at will, any given type of pond or lake, brook, or river. One of his masterpieces was a large aquarium which was a perfect model of an Alpine lake."

Konrad went to heroic efforts to protect and reclaim his animals after the war. To this day at Altenberg, at what is now the Lorenz Institute, you can meet his dog Stasi's descendants. The same loyalty extends to members of the original jackdaw and goose flocks, particularly to the children of Tchock and Martina. Though most pre-war geese were lost and did not overlap with the post-war flocks, the commitment to recording their biographies remained intense.

Yet no evidence exists that Lorenz ever inquired about his friend Bernhard as he hid from the Nazis in Holland. One letter does reveal that his wife Gretl discouraged Konrad from interceding; she didn't want to appear collaborationist. Another friend from Konigsberg (Dr. Baumgarten) tried to find help for Hellman, but to no avail. In the entirety of Lorenz's writing, Bernhard's

Bird Lovers, Backyard
THALIA FIELD

life and fate warrant significantly less mention than a long scene of Stasi, faithfully lying at Konrad's feet wondering, "Are you ever going to take me out?"

Reading science as biography or poetry feels both rich and problematic.

Lorenz: "When I look back now and remember how much we learned from our two ducks, they almost seem to be my most influential teachers. We took it for granted from the beginning that the ducklings would direct toward us the behavior patterns they normally displayed toward their natural mother. We were not in the least surprised to have them follow us everywhere."

In *The Jungle Book*, for their crime of resisting assimilation, the boy Mowgli leads the dholes (endangered red dogs from South Asia) into a hole where he and his "law abiding" wolf friends slaughter every one of them.

Social Organization without Love

Lorenz: "In the case of wild geese, I have repeatedly noticed that a betrothal was pledged when two fairly close friends met again after a fairly long separation. Even I myself have been affected by this quite typical phenomenon—but that is another story."

Recall that in Freud's narrative theory, catharsis or abreaction in one's storytelling purges, expresses, or discharges built-up emotional memory-content that was in tension or repressed. Readers seek pleasure in the release of psychic identification with fictional worlds experienced vicariously without shame or self-reproach. Even more, people see family dynamics as internalized representations of the political world. Conversely, what happens in politics stirs up fantasies of the family drama, which can provide impulse to action.

Lorenz: "I consider early childhood events as most essential to a man's scientific and philosophical development."

Do childhood stories train us?

Lorenz: "When in Kipling's Mowgli love is awakened, this all-powerful urge forces him to leave his wolf brothers and to return to the human family. This poetical assumption is scientifically correct. We have good reason to believe that in human beings—as in most mammals—the potential object of sexual love

makes itself evident by characters which speak to the depth of age-old inheritance, and not by signs recognizable by experience—as evidently is the case in many birds. Birds reared in isolation from their kind do not generally know which species they belong to... This phenomenon can be observed regularly in hand-reared male house sparrows, who, for this reason, enjoyed great popularity among the loose-living ladies of Roman society, and whom Catullus has immortalized by his little poem, 'Passer mortuus est meae puellae.'"

Fiction to biology to birds, back to people, back to poetry—authority lies in how you thread the sequence, solicit association. To drape one's science in the authority of the poets seems as common as poets using science to dramatize daydreams.

Lorenz: "Although, in that summer of 1909, we felt that we had outgrown the game of pretending to be ducks, we accepted our roles as duck mothers with passion and devotion."

On Aggression (the "So-Called Evil")

The experimental work that young Bernhard Hellman did with cichlid fish (putting a mirror in their tanks to demonstrate that they would fight their own reflected image to exhaustion) prompted him to coin the term *action-specific potentiality*, and identify the effects of dammed instincts. Bernhard's work on cichlids became a subject of Lorenz's research after the war.

In 1940, Lorenz wrote: "Precisely in the large field of instinctive behavior, humans and animals can be directly compared...We confidently venture to predict that these studies will be fruitful for both theoretical as well as practical concerns of race policy."

In *On Aggression*, Lorenz gives a rare anecdote directly about humans, to "prove" that violence, like the starling grabbing the invisible fly, expresses itself as a *vacuum activity*, an instinct with no relation to the world around it.

Lorenz: "So-called polar disease, also known as expedition choler (anger), attacks small groups of men who are completely dependent on one another and are thus prevented from quarreling with strangers or people outside their own circle of friends. From this it will be clear that the damming up of aggression will be all the more dangerous, the better the members of the group know, understand, and like each other."

Observations, poetry; it's a live, live world—crabs who cannibalize each other, seahorse fathers rearing the young—in the joy of recognizing sentience in other creatures, we feel the rush of kinship, and perhaps because we don't know how to act around lost or abandoned or ultimate kinship, we try to own other creatures' actions, translate them into our language, extend them our fantasies. We are industrious ants or hermaphrodite barnacles. We are fierce lions or clever foxes. We are monogamous penguins or self-sacrificing helpmates at the nest. We like these stories because it's hard to get a grip on exactly where we stand. No matter how many airplanes we build or satellites guide us, we feel like we're everywhere and nowhere, lost in our family without a map. We are all the animals and none of them. It is so often said that poetry and science both seek truth, but perhaps they both seek hedges against it.

In their famous pre-war collaboration, Lorenz and Tinbergen took the mother goose's egg from her and watched her rescue it by rolling it gently back to the nest. They took the mother's egg from her and watched her rescue it by gently rolling it back toward the nest and then they took it again. This time they never gave it back. They watched her try to continue the rescue by gently rolling the missing egg back to the nest.

These experiments are done on the whole animal in forced conditions. In these ways they mothered brood upon brood of mothers. What can friends and false mothers teach us?

From *The Jungle Book*:

I have taught thee all the Law of the Jungle for all the Peoples of the Jungle—except the Monkey Folk who live in the trees. They have no Law. They are outcasts...We do not drink where the monkeys drink; we do not go where the monkeys go; we do not hunt where they hunt; we do not die where they die. Hast thou ever heard me speak of the Bandar-log till today?...The Jungle People put them out of their mouths and out of their minds. They are a very many, evil, dirty, shameless, and they desire, if they have any fixed desire, to be noticed by the Jungle People. But we do NOT notice them even when they throw nuts and filth on our heads.

> Lorenz: "In reading those books, one feels that if an experienced, old wild goose or a wise black panther could talk, they would say exactly the things which Selma Lagerlof's Akka or Rudyard Kipling's Bagheera say." (Disturbance of Characteristic Behaviors through Domestication)

Digging in various records confirms that when they were teenagers, before Lorenz recieved funding for his first experiments, he relied exclusively on Bernhard Hellman's tanks of cladocera and cichlids. Bernhard, the better student, redacted books for Konrad, and introduced him to scientists he hadn't heard of. Instead of continuing his experiments, however, Hellman spent the early years of the war forced to make jewelry.

Sitting on the shores of his pond, watching his birds, Lorenz made the 'discovery' that a non-hybridized person would have intact aesthetic instincts. He summarized these 'findings' in his paper, *Breakdowns in the Instinctive Behavior of Domestic Animals and their Social-Psychological Meanings.*

> Lorenz: "I believe man has an inborn abhorrence for humans who have degenerate instincts. This abhorrence has also certainly a species-preserving value, since in humans degenerate mating drives and similar brood-care reactions go along with each other, as, e.g. with my greylag/domestic goose crosses."

What does someone hearing analogies hear? One thing makes another thing appear true in a new way—or reinforces an unstated suspicion. Perhaps a little "Aha!" results—the satisfaction of concept-making. A complex dance is initiated and even if the hearer later thinks, 'I shouldn't have gone along with that dance,' there was still an intimacy for a moment. Why are certain metaphors successful? Why is someone a star? A pig?

> Lorenz: "When we speak of falling in love, of friendship, personal enmity or jealousy in these or other animals, we are not guilty of anthropomorphism. These terms refer to functionally-determined concepts, just as do the terms legs, wings, eyes and the names used for other bodily structures that have evolved independently in different phyla or animals. No one uses quotation marks when speaking or writing about the eyes or the legs of an insect or crab, nor do we when discussing analogous behavior patterns." (Analogy as a Source of Knowledge)

Is textual action (what we're willing to do with words) where behavioral theory might begin?

> Lorenz: "Nothing is more important for the health of an entire people (volk) than the elimination of invirent types, which, with the most dangerous and extreme virulence, threaten to penetrate the body of a people like the cells of a malignant tumor."

Civilized Man's Eight Deadly Sins

Biology, beauty, and storytelling intimately align. "The true observer" is a rare and delicate man, says Konrad. Illustrating his own books, Lorenz sketched domestic animals and degenerate people. "Believe me, I am not mistakenly assigning human properties to animals: on the contrary, I am showing you what an enormous animal inheritance remains in man to this day." In a letter written in 1939, Lorenz refers to the "ugly Jewish nose" of a shoveler duck.

How can something that can be true in one story be a lie in another?

> Lorenz: *"As long as a tribe or a volk possesses a very high degree of racial uniformity, assessing an individual by his external characteristics alone will be possible and drawing inferences about the full value of his inner behavioral norms will be justified."*

In *King Solomon's Ring*, Lorenz mentions Bernhard Hellman in one sentence, while Gloria, Martina, Martin, Koka, Tschok, Stasi, and Tito get biographical chapters. To introduce the chapter titled "Martina," Lorenz explains, "Although the events I am about to relate took place fifty years ago, my notes are so reliable and my memory so vivid that I believe I can draw an informative picture of the life of a greylag goose by telling this story of my first goose."

> *"The wish [for an idealized past] makes use of an occasion in the present to construct, on the pattern of the past, a picture of the future."* (Freud, *Creative Writers and Day-dreaming*)

Can we imagine analogy, or even the feeling of sympathy between creatures, providing the source for knowledge—some uncontaminated epistemology—allowing people to share an animal's world? Definitely. But, one by one, analogies also reveal the mirror's opaque side, the confusion within this perceived agreement. In footnote and trunk song, stories give the illusion of completeness in hindsight. Lorenz knew that his support of race policy would have personal implications: "The fully superior person (Vollwertige) reacts against contemporaries manifesting inferior traits by keeping away from them."

Still, it's hard to see what analogy could be found for keeping away from your friends.

Analogy as a Source of Knowledge

> Lorenz: *"Believe me, one day everything I've said will seem very ordinary."*

What did all his stories about geese, jackdaws, dogs, and ducks tell about people?

> Lorenz: *"The inference was clear: I must quack like a mother mallard in order to make the little ducks run after me. No sooner said than done. When, one Whit-Saturday, a brood of purebred young mallards was due to hatch, I put the eggs in the incubator, took the babies, as soon as they were dry, under my personal care, and quacked for them the mother's call-note in my best Mallardese…For hours on end I kept it up, for half the day…In the interests of science I submitted myself literally for hours on end to this ordeal…I was congratulating myself on the obedience and exactitude with which my ducklings came waddling after me…"*

Perhaps there are no other creatures to turn to—ant, jellyfish, bird, or dog—to help explain us. Perhaps explanation isn't the point? Freud said writers allow readers to experience their daydreams without shame or self-censorship. The scientist may be no different. Perhaps tropes are merely features of his individual art. There is evidence of his writer's life but not so much his inner life. Is it good writing?

Our questions about Lorenz's life might not feel so insistent if he hadn't told so many animal stories. But maybe in all his writing about all his animals he was simply asking—to any friend he'd ever had—"Here am I—Where are you?" ■

Thalia Field has published three collections with New Directions (Point and Line; Incarnate:Story Material; Bird Lovers, Backyard), a collaborative essay, A Prank of Georges, and a performance novel, ULULU (Clown Shrapnel). Her forthcoming novel, Experimental Animals, explores the situation of the early laboratory in France and its aesthetic counterparts.

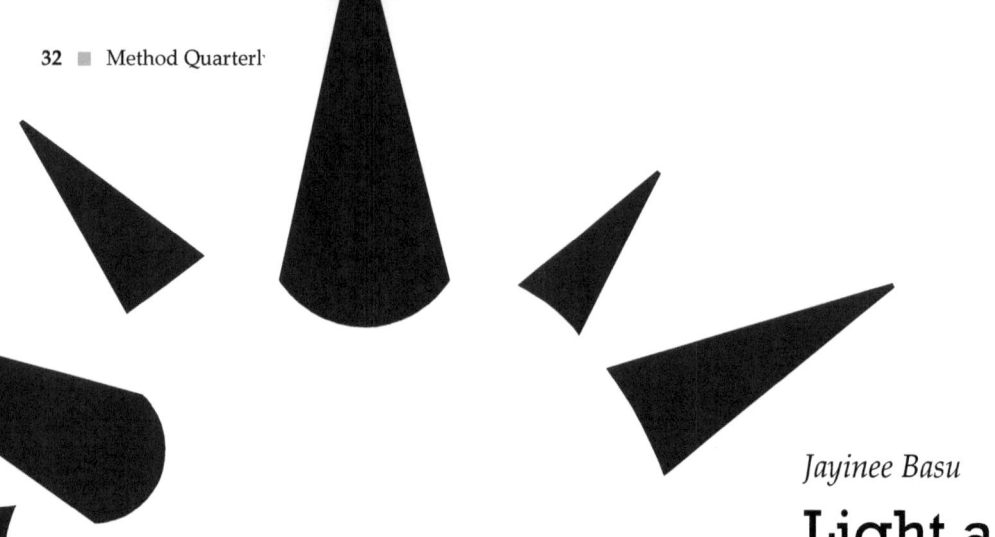

Jayinee Basu

Light and Madness

Imagine stepping into a botanical garden hothouse at night. Moonlight streams through the glass panes, illuminating creamy orchids and night-blooming cereus, glittering on the glossy dark green leaves. Your pupils dilate to take it all in. The eye allows us to absorb the tumultuous mess of visible light, the shapes and colors of our visual landscape. And yet when we look out into the world, instead of seeing nebulous whorls of stuff, we see petals, rivers, faces. We find the order in visual chaos by drawing boundaries around the disparate parts, grouping them together, and making them whole.

At the primary level of vision, our eyes find and exaggerate edges at the point at which they appear on the retina. Retinal ganglion cells act like an on-off switches, responding to the contrast between light and dark. Each of these cells receives input via separate tiny apertures, resulting in limited individual data that is useful only when integrated. Neighboring neurons inhibit each other; at the boundaries between two flat shades of grey, this inhibition emphasizes borders and contours, creating the ripples and highlighted peaks known as the Mach Band Effect.

Our minds also shape and exaggerate the boundaries that our eyes draw; how we think shapes what we see. When we think of a mushroom, we think of only the little cap and stem that we eat, not the massive rhizomorph that sustains the fungus across several underground square miles. In the first half of the twentieth century, the ecological psychologist J.J. Gibson was developing a relational theory of visual perception, based on this exchange between objects and thoughts. His affordance theory states that organisms perceive the world not only in terms of the shape and relationship between objects, but also based on how an object can be used. A branch can be differentiated from a tree because of its swingability, the mushroom from the mycelium because of its edibility. An affordance is not an inherent quality of any one thing, but lies in the relationship between object and subject.

Gibson's relational model of cognition was a radical departure from the psychological status quo of his time. Up until the 1950s, scientists understood perception as happening when the brain extracted meaning from the image of an object formed on the retina. To Gibson and other psychologist-philosophers, laboratory tests that relied on hyper-controlled and isolated stimuli could not properly model the variable and fluid nature of reality. Such closed-off experiments would lead to artificial conclusions about the nature of consciousness in a messy world.

R.C. JAMES, "DALMATIAN"

Likewise, the Gestalt psychologists used "biotic experiments" to extrapolate their laws of perception through real-world situations. The Gestalt laws define a theory of awareness as holistic instead of analytic: the whole is other than the sum of its parts. Like the sphere in the Gestalt image above, an object can be perceived even in a space where no sensory information is provided. Our brains fill in the gaps and we see things that aren't there.

R.C. James's image 'Dalmatian' famously demonstrates the Gestalt law of emergence. The viewer is presented with seemingly random splotches of ink that have no meaningful borders. Rather than identifying the pieces of the Dalmatian first, the shape of a Dalmatian sniffing the ground emerges all at once when the brain categorizes some of the shapes as belonging to the foreground and some as belonging to the background. It does this by looking at the proximity and similarity of certain shapes and folding them into the culturally acquired expectation that 1. there is something to see in the constructed image; 2. Dalmatians are spotted; 3. dogs sniff the ground, and so forth. The Dalmatian only comes into view when the image is seen concurrently as a whole, versus an additive process.

■ ■ ■

Back in the hothouse, you see moonlight glimmering in a black pond. Something is creating ripples and waves in the water, but it is too dark to see

exactly what. You squint your eyes toward the motion for more information. When the cells of your retina receive insufficient or ambiguous contrast information, the brain begins to apply perceptual boundaries to what is otherwise visually undifferentiated. We create meaning from ambiguity by noticing that even when an object or surface transforms (through an angle tilt, for example), it retains invariant properties: ratios of light intensity, gravity, the penumbra of shadow, or functional affordances. In the hothouse, you can approximate the location of the ripples based on the uniform rate of change in light as it radiates from the source. Soon the ripples die away and you look elsewhere, not having enough information to understand what caused the disturbance.

It's easy to imagine that the same brain processes that help us identify boundaries and name objects can also lead us astray, especially when information is restricted. The brain takes various shortcuts in order to conserve energy which can lead to strange conclusions. We can experience such shortcuts directly by observing the "Lilac Chaser" optical illusion.

In this animated illusion, a ring of lilac dots encircles a black cross, which we are instructed to focus on as an empty spot travels around the ring. Gestalt transforms the still images of the animated image into a moving picture. As you continue to stare at the cross, a secondary phenomenon manifests: the empty space becomes a green dot circling the lilac, rapidly eating them until all that is visible is the gray background underneath.

This intriguing secondary effect happens because the brain is lazy and tends to habituate to static stimuli. Staring at one spot means that the information presented to each neuron does not change. The brain adjusts to it rapidly, in the same way that you don't register the sensation of your clothes on your body seconds after you put them on, or how you stop noticing the smell of food cooking as you stand in the kitchen. Staring at the center makes the peripheral lilac dots disappear, coloring in the space with the surrounding grey. However, the circling green dot remains as an afterimage of the lilac. The resulting effect is very different from the actual stimulus.

Like in much of neuroscience, we've also learned a great deal about how thought and vision are connected by looking at cases in which they work differently. One of the most reliable symptoms to aid in the diagnosis of schizophrenia and other psychotic disorders is the reduced capability to fixate, or hold a steady gaze on a single unmoving object. People with schizophrenia are more likely to have trouble smoothly following the motion of a moving object, instead displaying a jerky back and forth motion of the eyes called a saccade. Saccades make integration of relative information about the environment more difficult, which may lead to divergent thoughts and incorrect conclusions. Even at an anatomical level, postmortem dissections show that individuals with schizophrenia have a 25% reduction of neurons in their visual cortex. The way we see influences the way we think.

Disruptions in integrating information can have major effects both visually and verbally. People with schizophrenia often describe their visual field as being fractured and disjointed, reporting that they only see parts of objects instead of the whole thing. Similarly, a hallmark of schizophrenic speech is "knight's move thinking," or sentences that don't follow a logical syntax, composed of thoughts and words that are only tangentially related. When objects seem isolated from another, the brain struggles to connect them, resulting in novel and oblique links.

This craving for associative meaning — and the resulting tendency toward paranoia — extends far beyond those with psychosis. Apophenia is finding meaning in meaninglessness. We are evolutionarily incentivized to come to quick conclusions about whether the motion of rustling grass is really a snake, so much so that sometimes we can really see the snake. While it's not particularly costly to be wrong about the snake and jump away for no reason, constantly seeing invisible snakes can make it very difficult to lead a normal life. Gestalt theory describes how we see what isn't there, helping us to make sense of the world. But these principles can also fail us; an over-reliance on these perceived boundaries creates a fractured picture of reality.

Today Gestalt theory has fallen out of style amongst cognitive psychologists. Without the rigorous controls of the prototypical lab experiment, Gestalt theory was unable to tease apart the specific neural mechanisms underlying perception, remaining descriptive rather than analytic. However, Gestalt principles maintain an important place in our understanding of cognitive function and dysfunction. They illustrate a progression from boundaries to objects to concept segmentation, culminating with integration into a full picture. They blur the boundary between the way we see and the way we think.

In that way however, the biotic experiments of Gestalt psychology can tell us something about the practice of science and the ways we see and think in the laboratory. Where we draw boundaries, define variables, and set controls shapes how we understand and interpret experiments. When dealing with complex systems like human consciousness, approaching the boundaries of truth may be the best we can hope for. You may never find out what the ripples in the pond are, but you may approach understanding when you see tiny dark purple blooms floating on the surface in the morning light. ■

Jayinee Basu is a writer based in San Francisco. A book of her poems entitled Asuras is forthcoming from Civil Coping Mechanisms.

NUCLEAR REGULATORY COMMISSION

Sarah Zhang

The Cat Went Over the Radioactive Mountain

Don't change color, kitty.
Keep your color, kitty.
Stay that midnight black.
The radiation that the change implies
can kill, and that's a fact.

Lyrics from "10,000-Year Earworm to Discourage Settlement Near Nuclear Waste Repositories"

To tell the mythology of Yucca Mountain, we might as well start with the fees. In 1983, a small fee of just a tenth of a penny per kilowatt-hour began appearing on electricity bills in America. The money was meant for Yucca Mountain, a wrinkle of land on the edge of the Nevada Test Site that was being turned into a massive tomb for the atomic age. Here, waste from nuclear power plants and weapons would be stored for at least 10,000 years until radioactivity faded to safe levels. Governments could fail and civilizations could fall, but Yucca Mountain was supposed to remain.

In 2014, after the Department of Energy had amassed $30 billion for the nuclear waste disposal fund, it quietly stopped collecting the fee. It stopped because a court told it to, because the Yucca Mountain Nuclear Waste Repository did not exist. Five miles of tunnels—out of the intended 40—had already been carved into the rock, but there was no radioactive waste stored there. After blowing past its planned opening date of January 31, 1998 by an embarrassing margin, the Obama administration in 2010 abandoned the languishing plans to build Yucca Mountain. Three and a half years later, a court ruled the federal government couldn't keep collecting fees for a site it had no intention of building. That's one way to see Yucca Mountain Nuclear Waste Repository's continued nonexistence, as yet another political boondoggle: thirty billion dollars of taxpayer money collected to build a mythical mountain .

But Yucca Mountain is more than that. The ambition behind it far exceeds the two- or four- or even six-year terms of any politician. Here we were trying to build a structure that would last longer than the Great Pyramids of Egypt, longer than any man-made structure, longer than any language. When forced to adopt a long view of human existence—when looking back on today from 10,000 years into the future—it's hard not to view Yucca Mountain in near-mythical terms. We can imagine future earthlings pondering it the way we ponder the Parthenon or Stonehenge today—massive structures imbued with an alien spirituality.

Ten thousand years may be the time scale of legends, but nuclear waste storage is a very real and practical problem for humans. It is a problem where incomprehensibly long time scales clash with human ones, where grand visions run up against forces utterly mundane and petty.

Radiation remains almost spooky, an invisible, silent, and odorless threat. In contaminated sites, you see men draped in full-body suits divining for radiation with Geiger counters. To someone who did not know the purpose of this, they might resemble robed members of an atomic priesthood appealing to some invisible power.

At high enough levels, radiation sears through the body, damaging tissue in ways that are immediately obvious. At low but still dangerous levels, you can't feel, hear, or see radiation passing through you, but it may knock loose a strand of DNA, creating a mutation that gets copied over and over in dividing cells until the day one of those cells becomes cancerous. It bides its time like a curse that can take years or decades to manifest.

In 1981, the Department of Energy convened a task force on how to communicate with the future.

The panel of consulted experts included engineers, but also an archeologist, a linguist, and an expert in nonverbal communication. Dubbed the Human Interference Task Force, they were tasked with figuring out how to keep future humans away from a deep geological repository of nuclear waste—like Yucca Mountain.

The repository would need some kind of physical marker that, foremost, could last 10,000 years, so the task force's report considers the relative merits of different materials like metal, concrete, and plastic. Yet the marker would also need to repel rather than attract humans—setting it apart from Stonehenge, the Great Pyramids, or any other monument that has remained standing for thousands of years. To do that, the marker would need warnings. But how do you

Sarah Zhang writes about science, technology, and culture from Berkeley, CA. She is a staff writer at Gizmodo.

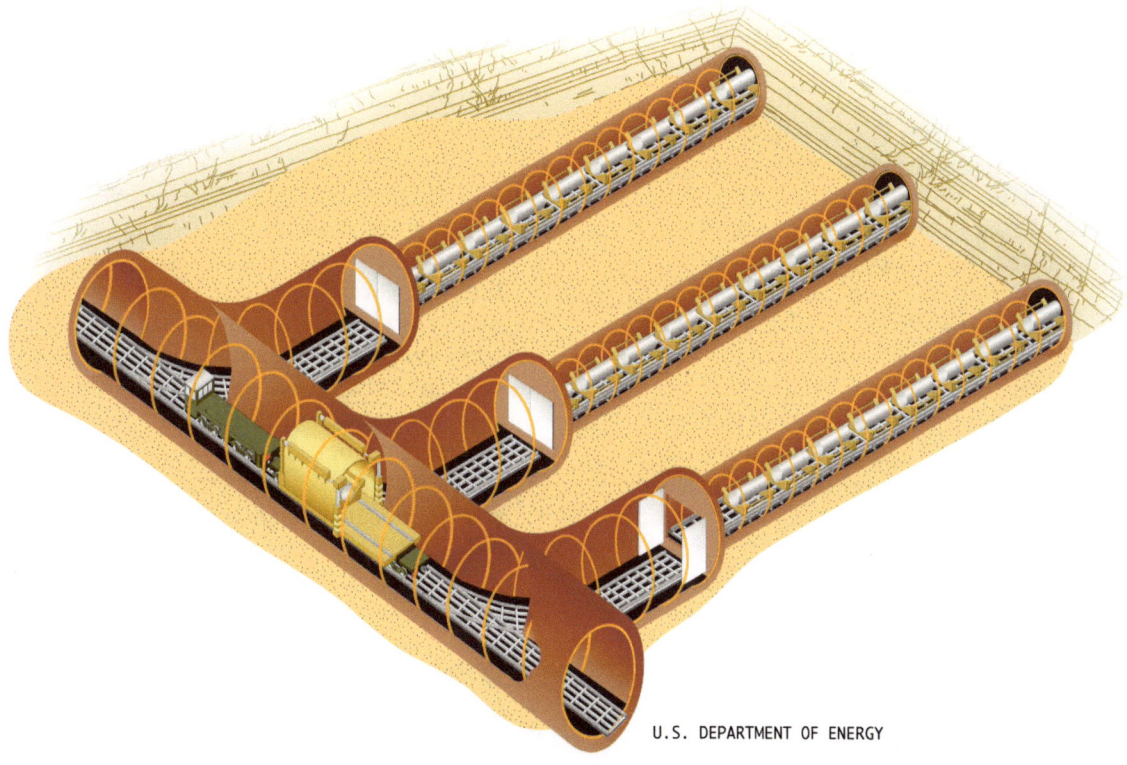

U.S. DEPARTMENT OF ENERGY

warn future humans whose cultures and languages will have evolved in unknown ways?

In addition to the physical marker, the task force recommends "oral transmission" to preserve their warning for future generations. Even as language itself mutates, the argument goes, the stories we tell endure. Imagine Homer's epics or *Beowulf*, but on an even longer time scale. The report, in characteristically dry language, imagines that the future population around Yucca Mountain might tell stories that "include perpetuation of knowledge about a 'special' place."

Thomas Sebeok, the linguist consulted by the Human Interference Task Force, goes into further detail in a separate report. He proposes seeding and nurturing a body of folklore around Yucca Mountain, even inventing annual rituals where these stories could be retold. These folktales need not explain the science of radiation; they simply need to hint at a great danger.

"The actual 'truth' would be entrusted exclusively to—what we might call for dramatic emphasis—an 'atomic priesthood,'" Sebeok writes. This group, he says, would need to include "a commission of knowledgeable physicists, experts in radiation sickness, anthropologists, linguists, psychologists, semioticians, and whatever additional expertise may be called for now and in the future."

In the decades when Yucca Mountain was still under development, some of the most intense interest in the repository came from 800 miles up the West Coast in Hanford, Washington. The Hanford Nuclear Reservation produced nearly all of the plutonium that went into the U.S.'s nuclear arsenal during the Cold War. Then, it was decommissioned. Now, it is the site of the largest environmental cleanup project in the country.

Fifty-six million gallons of radioactive waste sit in 177 steel tanks buried underground. The waste ranges from soupy to sludgy, and it has the unfortunate habit of leaking out of the aging tanks into the groundwater.

This wasn't the plan, of course. The idea was to build a vitrification plant on site, where radioactive waste could be mixed with molten glass and poured into steel columns—making the impermeable nuclear coffins that would then be entombed in Yucca Mountain. But the cleanup at Hanford has been horribly mismanaged. The vitrification plant, due to open in 2011, is still half complete. Of course, even if we manage to safely solidify and seal the radioactive waste at Hanford, we still don't have anywhere to put it.

Meanwhile, the radioactive waste keeps leaking.

Go back down the West Coast from Hanford, head inland for 200 miles , and you'll hit Carlsbad, New Mexico, home to the Waste Isolation Pilot Plant (WIPP), a nuclear waste repository in the deserts of the Southwest that was actually built.

Unlike Yucca Mountain, though, WIPP is only designed to handle low-level waste. While it can lock away the stuff that has come in contact with radioactive material, it can't safely store the byproducts of nuclear reactors themselves. Gloves, tools, and other equipment used to handle plutonium and uranium are packed in drums, which are then stored in rooms tunneled into the natural salt deposit. Over time, the salt will ooze around the barrels, encasing the waste in a mineral tomb.

The problem of long-term waste storage at WIPP is real but still theoretical. It's only when WIPP is shut down and sealed off that the plan to warn away future humans will be set in motion. For now, the tentative design is a series of 25 foot-tall granite monuments engraved with warnings in seven languages.

But, like Yucca Mountain, WIPP has been the subject of many more fantastical and evocative proposals. In 1991, it too convened a multidisciplinary panel to study the problem of communicating with the future. The resulting report laid out proposals that ranged from an architect's "landscape of thorns" to a warning message that begins like this:

> *This place is a message...and part of a system of messages...pay attention to it!*
> *Sending this message was important to us. We considered ourselves to be a powerful culture.*

U.S DEPARTMENT OF ENERGY

This place is not a place of honor...no highly esteemed deed is commemorated here...nothing valued is here.

What is here is dangerous and repulsive to us. This message is a warning about danger.

In February, a drum of radioactive waste at WIPP ruptured underground causing radioactive material to snake its way up a ventilation shaft and expose 21 workers on the surface 2,000 feet above the drum. WIPP has since shut down and may not reopen for years.

With Yucca Mountain gone, the Department of Energy had actually considered sending Hanford's to-be-vitrified high-level waste down to WIPP. This plan clearly wasn't going to happen either.

How can a drum just rupture? The official investigation points to a chemical reaction between nitric acid and trace metals in the drum. But this reaction only happens at high temperatures, which has cast suspicion on one other component in the drums: kitty litter.

Kitty litter is routinely used to help stabilize radioactive waste, but a contractor had recently switched from using a plastic-based litter to a wheat-based one. The rotting wheat may have created just enough heat to set off the chemical reaction that ruptured the drum.

In 1984, the German journal *Zeitschrift für Semiotik* (Journal of Semiotics) published a dozen responses from academics speculating on how to communicate across 10,000 years. The proposals range from the mundane to the bizarre and fantastical. One respondent proposes making the storage barrels impossible to open without great technical skill. Another involves creating a series of warnings in concentric circles that would expand as languages evolve. Thomas Sebeok is in there, elaborating on his system of rituals around an atomic priesthood and their rituals. But a pair of semioticians, Françoise Bastide and Paolo Fabbri, take the germ of Sebeok's idea for a nuclear folklore to a singularly strange conclusion.

Their solution is "ray cats," creatures bred to change color in the presence of radiation—like walking, purring, yarn-chasing Geiger counters.

But this is just the first part of the proposal. Alongside the cats, Bastide and Fabbri propose that we invent a body of folklore, passed on through proverbs and myths to explain that when a cat changes color, you better run.

Still more wonderful and unexpected is that someone has taken Bastide and Fabbri at their word and actually written a song about ray cats. The podcast *99% Invisible* commissioned Berlin-based artist Chad Matheny, aka Emperor X, to compose a "10,000-Year Earworm to Discourage Settlement Near Nuclear Waste Repositories" for an episode on nuclear waste. The song, writes Matheny, had to be "so catchy and annoying that it might be handed down from generation to generation over a span of 10,000 years."

Ray cats may not exist. A ten thousand year nuclear waste repository at Yucca Mountain may not exist. But a song about them does.

There are no fantastical creatures at Hanford, but there are rabbits and pigeons and swallows and tumbleweeds. The security buffer around the operating Hanford Nuclear Reservation has since been converted into a national monument untouched by agriculture and development. It's a lovely place to go hiking.

But to Hanford's Biological Control Program, the wildlife are potential "biological radiological vectors," and therefore represent a huge nuisance. Rabbits, badgers, and gophers that somehow ingest leaked radioactive material can spread their radioactive poop across thousands of acres. The radioactive creatures have to be hunted down, and their poop safely cleaned up by people in suits. Even tiny termites and ants can unearth radioactive material.

And then there are tumbleweeds, whose taproots can reach 20 feet down to suck up buried radioactive waste. In the winter, those taproots wither, and it's off the tumbleweeds go, tumbling miles away with the wind. In 2010, Hanford had to chase down 30 radioactive weeds.

Someday, our nuclear waste might actually be sealed off in a mountain capped with a giant monument warning future humans 10,000 years into the future. But for now, tumbleweeds—that casual symbol of tedium—keep tumbling away with our intractable nuclear waste. ■

WELLCOME LIBRARY

Leah Ginnivan

The Dirty History of Doctor's Hands

In 1846, Ignaz Philipp Semmelweis, a sad-eyed, mustachioed young medical graduate, became chief resident of obstetrics at the Vienna General Hospital. Over the next two decades of his brief career, he became the "savior of mothers" and an enemy of the medical establishment, driven mad by his quest for the truth about hospital-acquired infections.

In the 1840s, teaching hospitals operated by trading free medical care in exchange for the opportunity to practice on poor people. At Semmelweis's hospital, two clinics ran side-by-side. One clinic trained midwives, and operated with a maternal mortality rate of about one in 25. But in the other clinic, which taught medical students, one in ten women admitted would die before she left the hospital—in some months nearly a third of women died. The leading cause of death was childbed (or puerperal) fever, and both clinics were referred to in official papers as "houses of death." But, Semmelweis noted, mortality rates were perplexingly lower among women who insisted on giving birth in the streets and fields of Vienna rather than risk setting foot in the maternity clinics:

> *To me, it appeared logical that patients who experienced street births would become ill at least as frequently as those who delivered in the clinic. What protected those who delivered outside the clinic from these destructive unknown endemic influences?*

Despondent, Semmelweis left for Venice:

> *I hoped the Venetian art treasures would revive my mind and spirit, which had been so seriously affected by my experiences in the maternity hospital.*

Leah Ginnivan is a public policy researcher and campaigner with an appreciation of arcane Wikipedia.

But while the art treasures of Venice may have temporarily soothed his spirit, upon his return Semmelweis learned that his close friend, another doctor, had been pricked with a clumsy students' scalpel in the middle of an autopsy. The doctor quickly became ill and died. Semmelweis wrote:

> *Day and night I was haunted by the image of [his] disease and was forced to recognize, ever more decisively, that the disease from which [he] died was identical to that from which so many maternity patients died.*

In these pre-Louis Pasteur days, the medical establishment didn't yet know about bacteria. The main preoccupation of the time was with humours and miasmas, and 'treatment' for childbed fever involved inducing vomiting, bloodletting, blistering agents applied to the women's inner thighs, enemas, and liberal use of leeches with the aim of purging the fever heat from the body.

His friend's death—so similar to the deaths in the medical student clinic—led Semmelweis to hypothesize that the trainee doctors were exposed to 'cadaverous particles' in the course of the autopsies they conducted, which they then transferred to the new mothers. The midwives in the neighboring clinic, who concerned themselves only with births, weren't exposed to these cadaverous vapors. Semmelweis proposed, for the first time in medical history, a connection between touching cadavers and a risk of infection.

Semmelweis decided to act on his hunch. He instituted a clinic-wide policy of mandatory hand washing between cutting up a body and assisting in a birth. "[Hospital staff] had frequent opportunity to contact cadavers. Ordinary washing with soap is not sufficient to remove all adhering cadaverous particles. This is proven by the cadaverous smell the hands retain," Semmelweis wrote, introducing a chlorinated lime solution to the hospital. It removed the smell of death and would, hopefully, remove the particles too.

In the first three months, death rates plummeted from one in ten to one in a hundred. Semmelweis had shown that he could conquer childbed fever with handwashing.

Find more Semmelweis memorial stamps in "Medicine in stamps – Ignaz Semmelweis and Puerperal Fever" by A.D. Ataman et al., Journal of the Turkish-German Gynecology Association

■ ■ ■

Or so you'd think. In fact, Semmelweis' arguments were completely rejected by the medical establishment at the time. Because it came decades before Pasteur's germ theory of disease, and because Semmelweis himself could provide no theoretical explanation for his observations, his empirically-minded colleagues argued that Semmelweis seemed to be "reverting to the speculative theories of earlier decades that were so repugnant to his positivist contemporaries."

Danish physician Carl Edvard Marius Levy wrote a vitriolic attack of Semmelweis's findings. First, he deferred to the statistics, blaming the success of Semmelweis's hypothesis on normal fluctuations of mortality rates in maternity clinics. Next, he argued against the assertion that anything so small as to be invisible could cause death, arguing that, "With due respect for the cleanliness of the Viennese students, it seems improbable that enough infective matter or vapor could be secluded around the fingernails to kill a patient." Lastly, he questioned Semmelweis's methods altogether: why didn't he run a simpler experiment, by fully separating those working with cadavers from those aiding in births?

To be fair, Semmelweis did have major faults in his reasoning—he thought that *only* cadaverous particles caused the fever, and couldn't explain why some women still contracted fever in the midwife clinic. But, perhaps most importantly, his theories presented a behavioral conundrum for his fellow physicians: testing his hypothesis further could implicate them as dealers of death. Accusing doctors of haplessly *causing* disease was a slur on the gentlemanly art of medical practice.

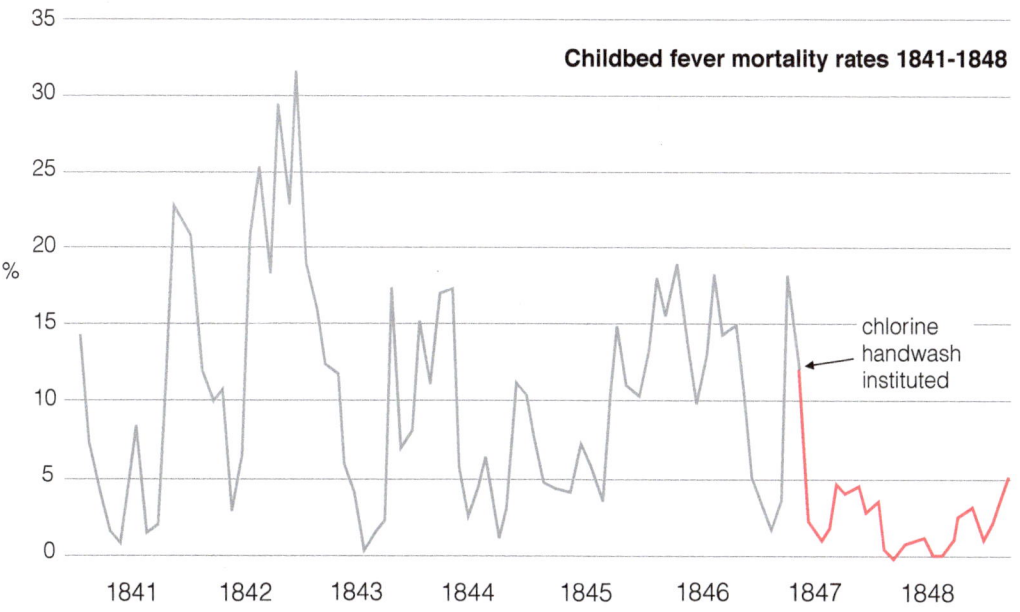

Such resistance within the medical community was short-lived. In the later part of the 19th century, a rising tide of empiricism—in particular the work of Joseph Lister and Louis Pasteur—lifted the lonely little ship of hygiene, and the benefits of hand washing would become universally accepted.

But while medical knowledge has radically shifted in the last 170 years, the reluctance to wash hands has persisted. The figures are striking: one large study

Transkei 45c
IGNAZ SEMMELWEIS
HOSPITAL MIASMA RECOGNIZED PUERPERAL FEVER AS A BLOOD-POISONING OR SEPTICAEMIA
JOHAN VAN NIEKERK   B2.4   1992

SEMMELWEIS MEMORIAL STAMP, ATAMAN ET AL.

found that hand washing rates were at just 26% in intensive care units, and 36% in the other wards (after monitoring systems were put in place, they jumped to about 50%). Another found doctors self-reported hand washing 73% o the time, but actually only did it 10% of the times they should have. The results are clear: doctors in hospitals not washing their hands kills roughly 100,000 Americans every year and sickens 1.7 million more.

So after 150 years, why is this still a problem? Excuses from doctors range from being too busy, to the washing solution and alcohol rubs drying out their hands, to constantly having to carry equipment which makes it difficult to wash, to hand washing facilities being inconveniently located. Others simply say they forget.

Today, a whole industry of high-tech ways to remind doctors to wash their hands has sprung up, selling hospitals vibrating sensors to remind doctors to lather up, intensive video-monitoring, and incentive schemes for good hygienic practices. One hospital introduced waist-high monitoring that buzzed when doctors walked past without washing their hands. But doctors got down on their knees and crawled under the sensors, just to avoid washing their hands. While busy doctors may occasionally be absent-minded, such reports seem to indicate that sometimes they knowingly avoid it.

Several studies have shown that nurses wash their hands more than doctors. Ironically, part of this may go back to the resistance

among Semmelweis's 19th century peers: It's speculated that doctors develop a complex of invulnerability—that, as medical professionals, they can't be harmed or harm others. "The ego can kick in after you have been in practice a while," one emergency department physician told the New York Times. "You say: 'Hey, I couldn't be carrying the bad bugs. It's the other hospital personnel.'"

Doctors are pretty much universally regarded as empirically-minded. But knowledge doesn't always translate to universal shifts in behavior—regardless of how easy, necessary, or cheap the solution. The clean, elegant answers produced by biomedical science can't be found in equal measure in the dirty world of human actions and motivation.

And what about our hero, Dr. Semmelweis? Increasingly obsessive over the years, Semmelweis took his colleagues' reluctance to accept his theories as a personal affront. Rather than launch a charm offensive to win his fellow obstetricians over, he wrote them abusive public letters: "I declare before God that you are a murderer and [history] would not be too unfair if it remembers you as a medical Nero," he told one. Increasingly driven mad by the world's failure to appreciate the importance of hand washing, likely exacerbated by a a touch of syphilis or Alzheimer's, his colleagues eventually had enough: three obstetricians signed referrals committing him to a mental asylum. One day, while on vacation with his wife and child, he was met at a train station by an old friend who wanted to show him his sanitarium.

Under this pretext, the 47-year old Semmelweis was driven straight to a large, public asylum. He was severely beaten by the guards and died—of an infection—two weeks later.

In honor of Semmelweis's legacy to medicine, several medical schools, hospitals, womens' clinics, and museums now stand proudly bearing his name. But, perhaps most appropriately, his name graces the so-called "Semmelweis reflex": the kneejerk reflex to reject new evidence contradicting established norms. ■

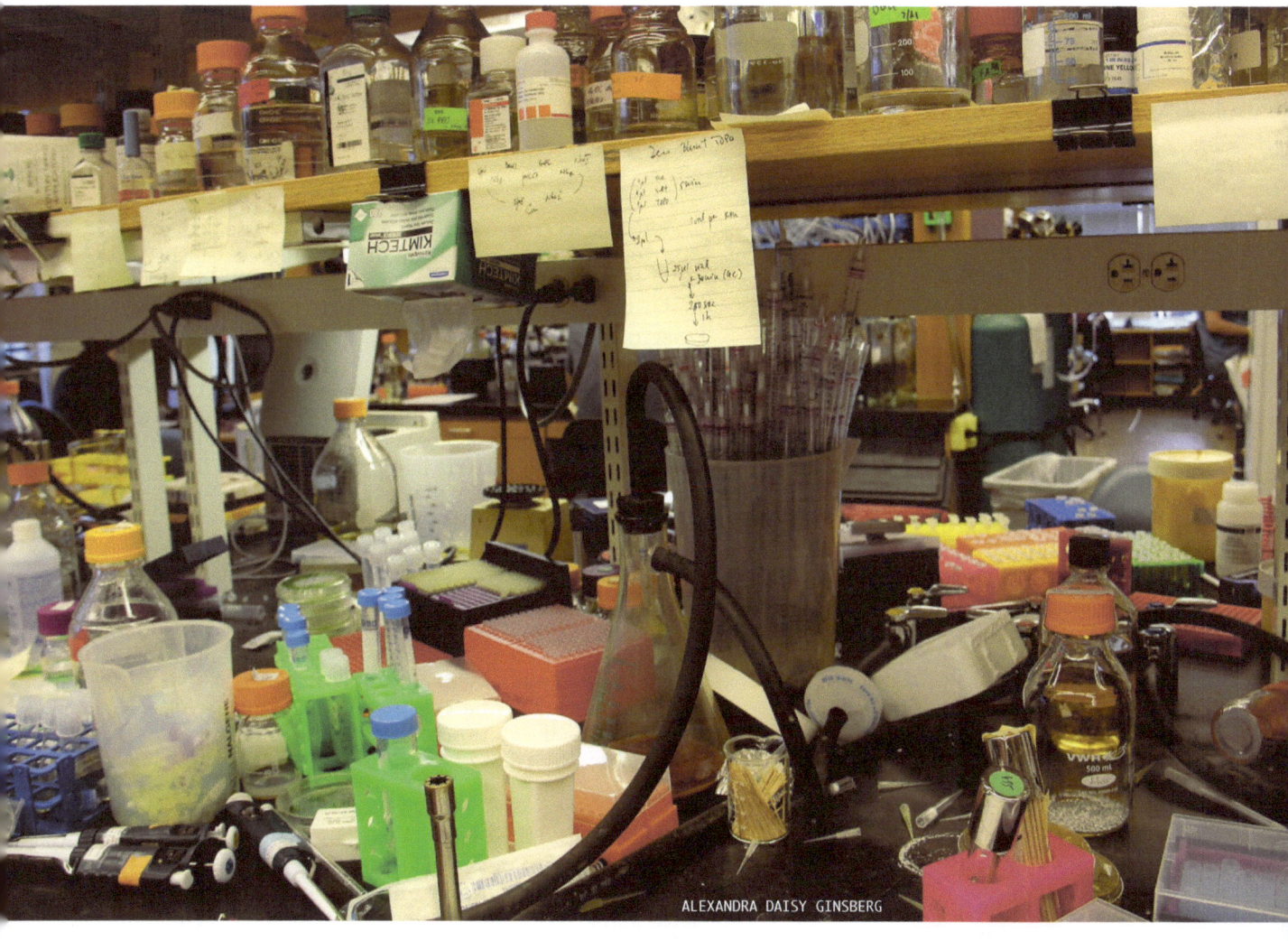

ALEXANDRA DAISY GINSBERG

Devin Burrill

Protocol: DNA Extraction

I start nearly all of my experiments with the same procedure: extracting recombinant DNA from *E. coli*. In my research, I alter DNA sequences in order to create drugs that target specific types of cells in the human body. As a biological engineer, I stitch pieces of genes into circular pieces of DNA (plasmids) to create new cellular pathways. Though many of the protocols I use in the lab take a long time and have a high rate of failure, DNA extraction is simple, works 99% of the time, and takes less than 30 minutes.

Creating a new plasmid is an iterative process. As I add each new piece to my plasmid, I move the DNA in and out of *E. coli* cells. The cells act both as living storage and tiny Xerox machines, reproducing the DNA and filling the cells with copies that I'll use in the next step of the process. I extract the DNA, cut and paste new genes into the plasmid, and insert it back into a fresh set of cells. Eventually I will harvest the complete plasmid from *E. coli* and transfer it into a yeast or animal cell.

Devin Burrill is a postdoctoral fellow at the Wyss Institute for Biologically Inspired Engineering at Harvard University. Her current research focuses on engineering targeted drugs for treating anemia, autoimmune disease, and viral infection.

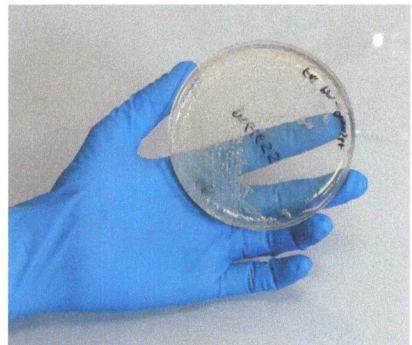

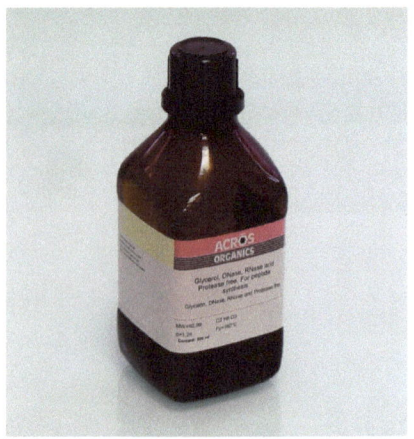

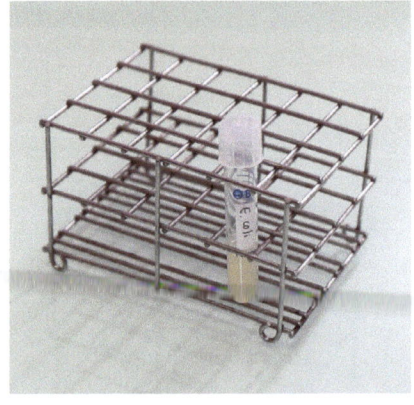

Below is a general protocol for extracting plasmid DNA from *E. coli* bacteria cells. The overall goal is to separate the desired plasmid from other cellular components (RNA, protein, chromosomal DNA, etc.).

1. I either start from a bacteria growing on a petri dish or from vials of bacteria stored at -80 degrees Celsius. *E. coli* containing recombinant plasmids are typically stored frozen in glycerol, which serves as a cryoprotectant ---- just like frozen food, we don't want our bacteria to get freezer burn! Using a sterile toothpick, I take a little bit of bacteria and put it in a test tube containing Lysogeny Broth (LB; often LB is referred to as Luria broth or Luria-Bertani broth, after the scientists who developed the medium). LB is a rich food for the bacteria that provides them with all the protein building blocks they need to grow.

2. *E. coli* are gut bacteria, so they grow best at a normal body temperature of 37°C. They also need to be kept shaking at around 200 rpm to prevent them from settling to the bottom and squishing each other. I put my tube of media with the few frozen cells in a shaking incubator and the cells start to grow. *E. coli* double every twenty minutes; the culture will become cloudy with floating cells within 4 to 12 hours.

3. Once the culture is cloudy, I need to separate the cells from the LB. I remove the test tube from the shaker and spin it in a centrifuge until all the cells arev pelleted at the bottom of the tube. The cells are all clumped together as a beige pellet at the bottom of the tube, and the LB above looks clear again. The media above the pellet is the supernatant and isn't necessary for future steps. I pour off the liquid in a waste container.

4. The bacteria next need to be resuspended in a liquid that provides the appropriate environment for busting open the cells and releasing the plasmid. Resuspension buffer is typically made at pH 8.0 and contains a buffering agent to help maintain the pH, RNAse to remove any of the cell's RNA that could get mixed up with the plasmid, and a chemical called EDTA, which absorbs any ions released during lysis that could harm the plasmid DNA. I pipette the cells up and down in the buffer, resuspending the bacterial pellet in enough buffer such that no cell clumps remain.

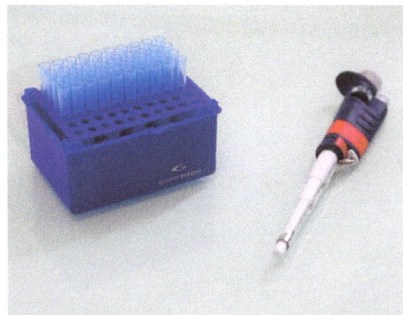

5. After the cells are completely resuspended, I'm ready to bust them open and retrieve the plasmid DNA. This process is called cell lysis: λύση is a Greek word that means "to set free" or "to release." The lysis buffer is typically alkaline (pH 12.0-12.5), to aid in denaturing chromosomal DNA and protein, while allowing plasmid DNA to remain stable. The two most important ingredients of lysis buffer are detergent (typically sodium dodecyl sulfate) and sodium hydroxide. Detergent breaks open the cell membrane, causing the cells to burst and release their contents. Sodium hydroxide helps denature chromosomal DNA, plasmid DNA, and cellular proteins.

After adding lysis buffer to the resuspended cells, I gently shake the tube to mix the contents. I see the mixture turning into something that looks like a glob of snot.

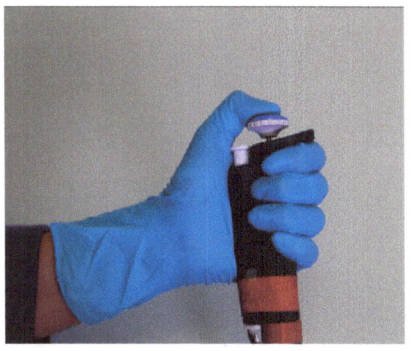

6. Next, I add a neutralization buffer—typically a concentratedsalt solution such as acidic potassium acetate—to the tube and shake to mix thoroughly. The high salt concentration causes any detergents to precipitate from solution, forming salt-detergent complexes that trap the stuff I don't want: proteins, chromosomal DNA, and other cellular debris. Under these conditions, plasmid DNA will re-fold properly and remain soluble—to me this is the most magical part!

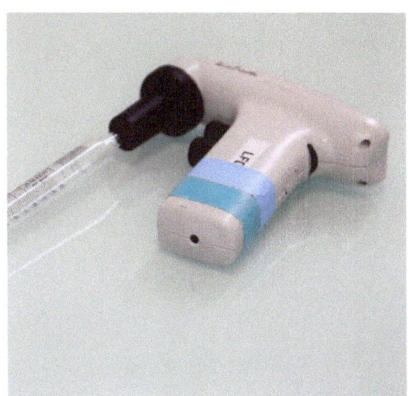

7. I spin the tube in a centrifuge to pellet all of the junk I don't want to the bottom. In this case, the supernatant contains what I want: the plasmid DNA in solution. I use a pipette tip to remove the clear supernatant and transfer it to a clean test tube.

8. I add isopropyl alcohol to the supernatant and shake the tube to mix thoroughly. Because DNA is not soluble in isopropayl alcoohol, it comes out of solution and becomes visible. The DNA looks like a floating white tangle of threads that some scientists call the medusa, because it looks similar to Medusa's head of snakes. Once I see the medusa, I know that I have successfully extracted my recombinant DNA. I centrifuge the tube again to pellet the medusa to the bottom. The supernatant is now just the buffers and isopropyl alcohol, so I pour it off and get rid of it.

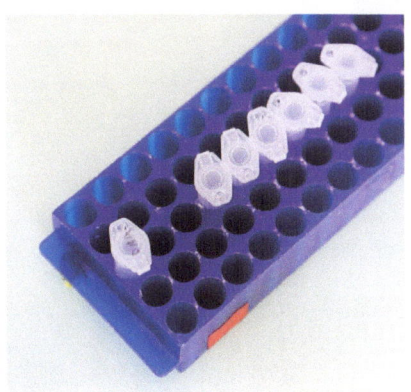

9. Once the liquid is removed, the DNA can be washed in ethanol to remove any remaining buffers or other residue from the extraction procedure. Once the ethanol is poured off, the DNA needs to be air dried to remove all the alcohol, which would prevent it from going back into solution again. Once it's dry, I resuspend the DNA in buffer and store it in the refrigerator or freezer, ready for my next experiment. ■

PHOTOS: FANNY DOUARCHE

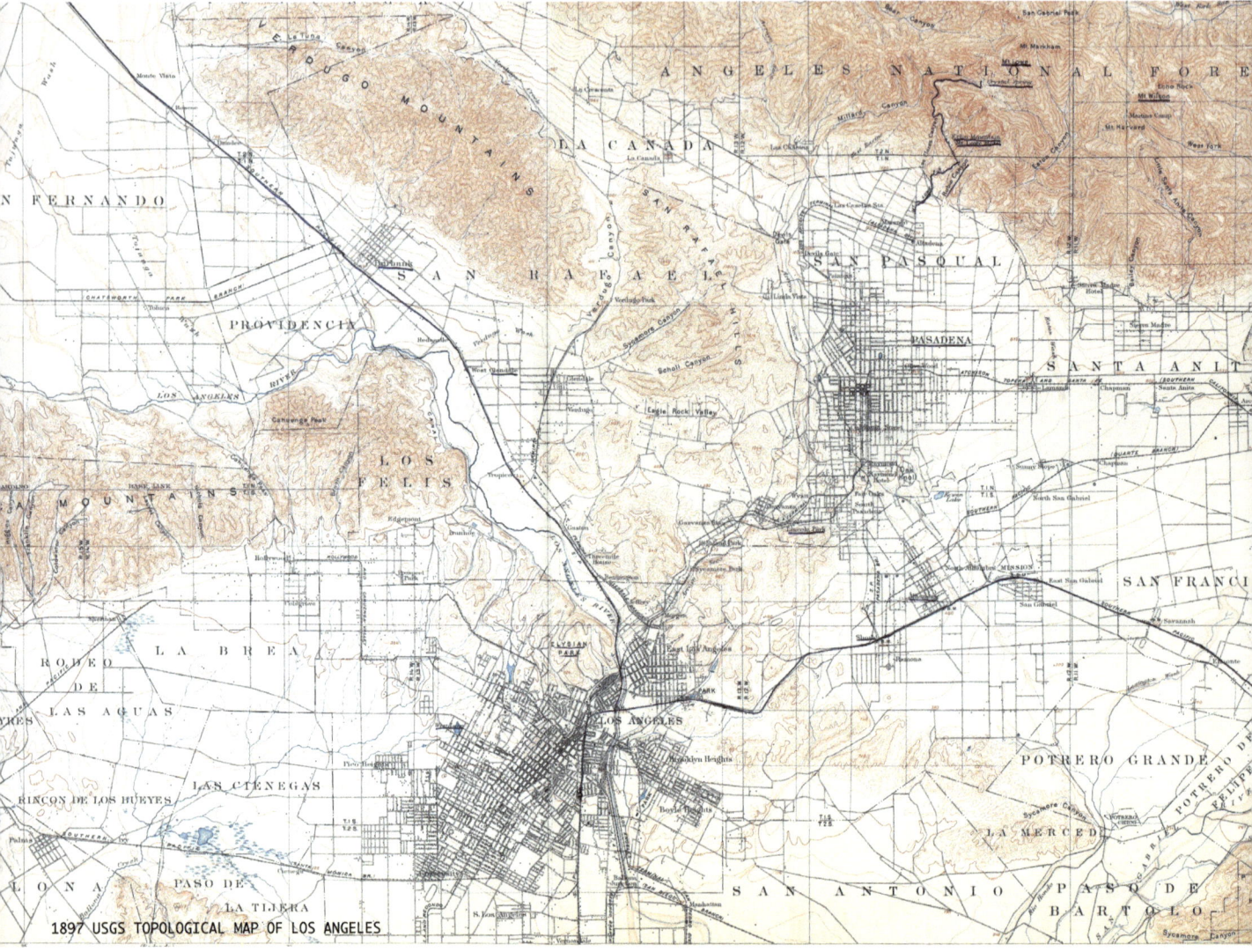

1897 USGS TOPOLOGICAL MAP OF LOS ANGELES

Laura Bliss

Coyotes Among the Seals

Many visitors to Seal Beach, California think it gets its name from the 5000-acre Navy depot that employs so many of its residents. But they're wrong; it's named for the animal that once colonized the sandspits near the beach. In the 1900s, during Seal Beach's short life as a resort town, vacation-goers could stand on the pier and watch thousands of the glistening mammals flop and moan, wriggle in and out of the water, out of the deep of the sea and onto the spit. Town boosters spread word of Seal Beach using imagery of the animals: seals dancing with a ball perched on the nose, seals serving up a gourmet lunch, seals cranking back in their office chairs, smoking cigars, the riches of their profitable appeal. Come to Seal Beach, they said, watch us at arm's length, as a destination scintillates before you: plinking music, a casino, a rollercoaster from the San Francisco World's Fair.

But two wars and the Depression quickly revised the Seal Beach fantasy. The pier disappeared, the Navy depot was built, and offshore drilling began in the mid-1940s. Sometime around then the seals checked out, leaving a town with a misled sense of identity and, perhaps, relationship to its wildlife. For this is the same place, just south of Los Angeles County, that passed an ordinance last month to exterminate all of its coyotes.

Coyotes' attacks on Seal Beach pets seem to be increasing. The "seem to be" is deliberate: It's not clear whether the coyotes' encounters with domesticated animals are in fact growing, and local authorities have admitted the difficulty of ascertaining firm numbers on either count. But what's certain is that as coyotes populate unremittingly across the U.S., and in Seal Beach, there's been a recent echo effect in anguished reports of coyote encounters. Seal Beach residents are gathering in front yards and stringing photographs of canine victims to the predators to branches, like so many tear-jerking Christmas trees. They are organizing

Facebook groups called "Coyote Watch," and "The coyotes are a problem for Seal Beach." They are calling community meetings to air stories of their traumatic encounters, like: a man who came home from a daughter's school play to find the pincers attacked and dying in the backyard. A woman who carries precautionary mace, a whistle, and a headlamp on dog walks around the neighborhood. Another who has seen a coyote regularly enough at her doorstep that she has named it "Joe" (normally naming an animal is a sign of affection, but apparently not in this case). "I don't want to co-exist with the coyote here in Seal Beach," one man told the Seal Beach City Council last August. "I want my daughters...and my wife to walk the dog through the neighborhood and feel safe—and we don't.

Fears crescendoed at that City Council meeting earlier this summer, and in September, councilmembers agreed unanimously that the solution to the coyote problem was to trap and gas the animals with carbon dioxide. It's a decision that's been met with considerable

media attention in the state; few trapping programs still exist in California since historically, trapping has backfired. Research shows that weaker coyotes tend to be those that fall prey, leaving the larger and more intelligent to mate and even increase a population. And in California animal shelters, gassing is an illegal method of euthanization. Activists have protested Seal Beach's plans as inhumane, organizing their own protests in support of more "integrated" measures to protect pets and humans alike: coyote 'hazing' (in an encounter with a coyote, create noise and try to appear as large as possible), keeping trash cans locked up and food inside, walking pets in daylight hours, and never leaving them alone in the yard.

Traumatized pro-trappers would probably argue that these measures don't do enough to address coyotes' brazen nature, and on one level they're probably right. It's entirely possible that a coyote might visit your yard while you're waving your arms and yelping in it, or that you'd encounter one on a morning jog, or that one might come not in search of your trash, but for innocuous fallen citrus from the trees in your yard.

KENT WANG

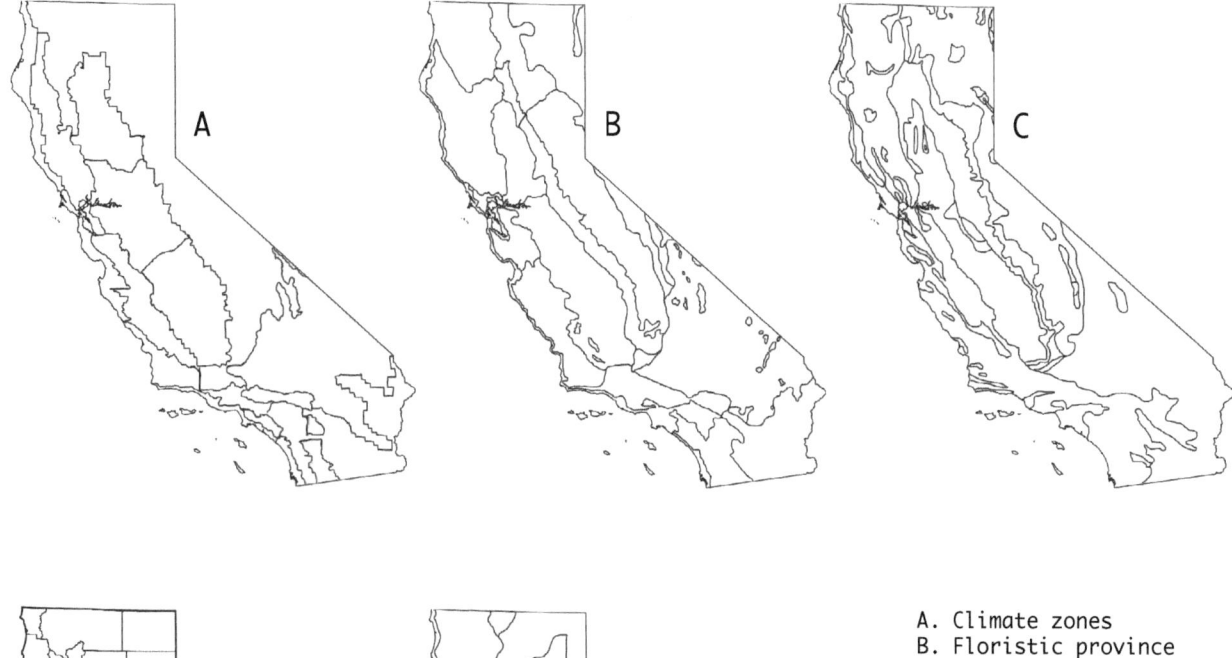

A. Climate zones
B. Floristic province
C. Agricultural zones
D. County lines
E. Interstate freeways

But Seal Beach residents may also suffer a chronic misunderstanding of the place they live. Southern California is a region defined by what biologists term the 'ecotone': a transition zone between ecological communities "whose boundaries may be relatively abrupt (as between chaparral and lawns) or gradual (as between domestic, feral, and wild animals)." That's Mike Davis writing in *Ecology of Fear*, a book that argues how the evolution of urban Southern California—more than any other citified region in the U.S.—cannot be understood apart from its precarious seat in nature. From its earthquakes, its seasonal fires, the drought that is the status quo, the defining edge of mountain and desert, it's a place whose basic contract rests on the interplay of 'urban' and 'wild.'

Coyotes are the tricksters of the Southern California ecotone; they pass cunningly across borders that are fuzzy to begin with, shifting shapes depending on conditions. Some hunt in packs, others individually. Some sleep at night, others are nocturnal. Take away one kind of food source, and coyotes will find another. Coyotes will eat fresh prey if they have the option—like squirrels, rabbits, mice, and sometimes deer—but they'll also eat berries and roots, human trash, fallen fruit, pet food, and pets, if they are desperate. (Some have attributed the apparent rise in coyote predation on small dogs and cats to diminished food supplies as a result of the West's historic drought).

It's no surprise that coyotes figure prominently in Native American literature, which at turns praises and admonishes the canine for its interminable curiosity and refusal to abide by the rules of any one animal or any one world. In this canon, the insatiable, brazen coyote regularly takes advantage of its knowledge of other beings, satisfying its hunger by manipulating the hunger of others.

Coyote was going along by a big river when he got very hungry. He built a trap of poplar poles and willow branches and set it in the water. "Salmon!" he called out. "Come into this trap." Soon a big salmon came along and swam into the chute of the trap and then flopped himself out on the bank where Coyote clubbed him to death. "I will find a nice place in the shade and broil this up," thought Coyote.

In his catalogue of rule-breakers mythical and real, *Trickster Makes This World*, Lewis Hyde writes of how the mythical Coyote takes advantage of his prey's instincts to capture them. In this case, a Nez Pierce tale, the salmon that swims upstream to spawn leaps into Coyote's trap out of natural hunger: "The worm just sits there; the fish catches itself." In another story from the Crow, Coyote traps two buffalo by coercing them to stampede towards the sun—where they can't see where they are going—then over a cliff.

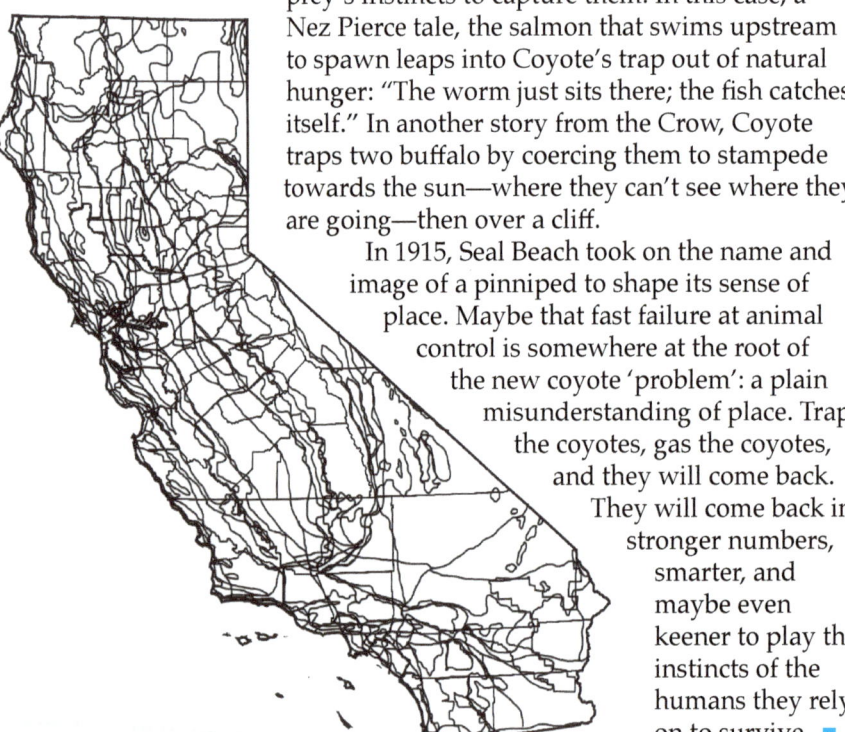

In 1915, Seal Beach took on the name and image of a pinniped to shape its sense of place. Maybe that fast failure at animal control is somewhere at the root of the new coyote 'problem': a plain misunderstanding of place. Trap the coyotes, gas the coyotes, and they will come back. They will come back in stronger numbers, smarter, and maybe even keener to play the instincts of the humans they rely on to survive. ■

Laura Bliss is a writer and journalist living in Washington, D.C.

Azeen Ghorayshi

Relics of Imagined Futures

"We are as gods and might as well get good at it."
– Stewart Brand, *Whole Earth Catalog* (1968)

"We are as gods and HAVE to get good at it."
– Stewart Brand, *Whole Earth Discipline* (2009)

WILLIAM HARTMAN

Everyone has heard of de-extinction. Everyone. If they haven't, and are brave enough to ask casually in conversation, and if the person responding doesn't want to take the time to actually explain it—or doesn't actually know—then the response will consist of exactly four words: *It's basically* Jurassic Park.

As a result, the story is everywhere. While some of my no less-obscure Google Alerts lie dormant for weeks before cropping up with new material from some weird corner of the internet, de-extinction merits a several-times-weekly type discussion. Everyone—from *Think Christian* to the *Martha's Vineyard Gazette*—is interested in the fact that this could really be happening. *Jurassic Park* could *really be happening.*

But is it? Yes and no. What has been framed within the Biblical rhetoric of revival and resurrection is really occurring using the now commonplace approaches of genetic engineering, amped up to a level that—while undoubtedly impressive—raises many more questions about what we can and should do with a technology that's already steeped in ambivalence.

"This is not about making the perfect wooly mammoth—it's about making cold-resistant elephants," said Harvard geneticist George Church at a conference in October. Church's lab is currently working to revive a wooly mammoth-like elephant in hopes of restoring the grasslands of the Arctic tundra. Or, as he told *New York Times Magazine* in response to criticism regarding the project's hype from fellow researchers, "I would like to have an elephant that likes the cold weather. Whether you call it a 'mammoth' or not, I don't care." (The *NYT Mag* article in question ran with the headline "The Mammoth Cometh.")

Techno-scientifically, engineering an elephant that is fat, hairy, and has cold-adapted hemoglobin would be an impressive feat requiring "a few dozen" changes to the animal's genome. But *resurrection*, this is not.

So why is it called de-extinction in the first place? The story goes back to Stewart Brand, a man who has done more to define our day-to-day experience in the digital age than most people who actively played a part in creating it. To put it simply, the 75-year old Brand loves playing with big ideas, and the idea of bringing back extinct species is nothing if not huge.

■ ■ ■

The *Whole Earth Catalog* has maybe one of the best origin tales, a relic as well-worn as the yellowing, wrinkled copies now propped up on used bookstores shelves across the United States.

Brand graduated from Stanford with a degree in ecology in 1960 before completing a short stint in the army and returning to San Francisco in the midst of its countercultural upheaval. In 1966, he began a public campaign to have NASA release a photograph of the whole Earth taken from space. Brand argued that seeing the planet in its entirety could be a powerful symbol to galvanize environmental stewardship from all who occupied "Spaceship Earth," a term popularized to emphasize the vulnerable nature of our tiny vessel's resources. In 1967, NASA took that photo. And in 1968, Brand founded what became one of the defining publications of the American counterculture, featuring the NASA-released photo and the words "WHOLE EARTH CATALOG / access to tools" on its stark black cover. Small-scale tools, Brand argued, were the keys to personal empowerment.

Over the next four years, the *Catalog* would come to sell over 2.5 million copies. While on one level it functioned as a purveyor of small-scale technologies to the hundreds of thousands of young commune dwellers scattered across the US, it also served as a point of communication for other communities that Brand, a seamless networker, moved fluidly between: most notably, the experimental art scenes in New York and San Francisco, the Bay Area psychedelic community, and the world of industry-based science and technology.

As such, the *Catalog* contained such seemingly incongruous items as buckskin jackets, Buckminster Fuller's geodesic domes, solar collectors, and wood stoves, as well as a $4900 Hewlett-Packard desktop calculator, subscriptions to magazines like *Scientific American*, and books on ecology and cybernetics. To contemporary readers of the *Whole Earth Catalog*, its first line, "We are as gods and might as well get good at it," rang the clarion call of a new sort of technological embrace.

"When these groups met in its pages, the *Catalog* became the single most visible publication in which the technological and intellectual output of industry and high science met the Eastern religion, acid mysticism, and communal social theory of the back-to-the-land movement," writes media theorist Fred Turner. Historian Andrew Kirk further argues that this iconoclastic forging of communities represented a "new alchemy" of environmentalism that emerged in the 1970s, where environmentally conscious consumption began to count as environmental awareness—or even activism—with technology serving as its main conduit.

In April 2013, over four decades after the final issue of the *Whole Earth Catalog* was published, a Berlin-based art group put together an exhibition entitled, "The Whole Earth: California and the Disappearance of the Outside," part of an ongoing series called "The Anthropocene Project." The catalog's stark white cover featured the image of the moon in the foreground, the earth tiny and vulnerable in the distance, with a large shadowed hand extending down towards it—either reaching out as peace offering, as acknowledgement, or threatening to scoop it up for its own use. The exhibit dealt with "the transgression of boundaries and the emerging consciousness of limits in the sign of the 'one earth,'" the curators wrote. "The *Whole Earth Catalog*… played a key role in mediating and popularizing images of the 'earth system.' The vision of the global Internet and central concepts of the ecology movement can trace their roots back to this moment."

■■■

In February of 2013, Brand gave an eighteen-minute TED Talk entitled, "The dawn of de-extinction. Are you ready?" In the talk, which has since been viewed over 1.6 million times and has been translated into 24 languages, Brand told a tale in broad strokes of the wildlife-destroying Anthropocene, as evidenced by the extinction of the once ubiquitous passenger pigeon.

"Sorrow, anger, mourning. Don't mourn—organize," Brand told an audience of people who had each paid several thousand dollars to attend the popular talk forum devoted to spreading the "power of ideas." "What if you could find out that, using the DNA in museum specimens, fossils maybe up to 200,000 years old could be used to bring species back?" Brand asked. Thanks to new developments in genome assembly, synthetic biology, and interspecies cloning, Brand said, bringing extinct species back into existence was now a distinct possibility.

This moment effectively served as the launching pad for Brand's newest venture, an organization called Revive & Restore that would be devoted to resurrecting extinct species using biotechnology. The project would be just one arm of Brand's Long Now Foundation, a non-profit dedicated to getting "people thinking past the mental barrier of an ever-shortening future." Perhaps most well known for its 300-foot-tall stainless steel 10,000 Year Clock—financed by a $42 million investment from Amazon.com founder Jeff Bezos and built inside a mountain that Bezos owns in Texas—Long Now's most recent project was sending a nickel disk inscribed with 1,500 languages to "Rosetta's Comet" via the Rosetta space probe. While these projects are at turns overly glorified or dismissed as New Age oddities, they are foremost large-scale art projects. Viewed in this way, the de-extinction project is the perfect addition to Brand's portfolio.

While Revive & Restore's flagship project is to clone the passenger pigeon back into existence, it is also aiming to resurrect the wooly mammoth in collaboration with Church's lab at Harvard. In addition, it plans to serve as a de-extinction hub of sorts, convening meetings and fostering communication across disciplinary boundaries among the scientists, conservationists, and ethicists working on relevant aspects of the project worldwide.

However, the project also has another, bigger goal: Brand wants to persuade the environmentalist and conservationist communities, which he repeatedly argues are stuck in a negative view of the world, to instead embrace the optimism of technology. "The environmental and conservation movements have mired themselves in a tragic view of life," Brand wrote in a letter to Church and biologist E.O. Wilson before launching the project. "The return of the passenger pigeon could shake them out of it—and invite them to embrace prudent biotechnology as a Green tool instead of menace in this century."

"Could be fun. Could improve things. It could, as they say, advance the story," he wrote.

A year later, however, the story has mostly raised lots of questions from disparate stakeholders staring at eachother across a vast expanse of muddled misunderstanding.

How arbitrary or specific are species boundaries in the first place?

In an era when endangered species are being cloned in zoos, where are conservation's limits?

What role have/do/should humans play in reinventing nature?

These questions are the beauty of a project whose grandiose packaging far outweighs its realities: Revive & Restore currently has only one full-time employee, a passenger pigeon-enthusiast with a bachelors degree in ecology but no advanced scientific qualifications. He has admitted it could be up to two decades before anything closely resembling a flock of passenger pigeons takes to the skies. Yet it's advancing a story.

In one of the few bioethical analyses of de-extinction written to date, author Ronald Sandler concluded by saying that, "The considerations in favor of de-extinction are largely techno-science oriented, not conservation-oriented." Given that there is no pressing need to pursue it, from Sandler's ethical perspective, "De-extinction is a luxury."

It may partially be an exercise in seeing what we can do with the latest genome editing technologies, whether we can make a creature as large an elephant give birth to something slightly different, whether we can then teach those animals to live and breed together, whether they can then manage to do so in the Arctic, and whether then some of the damage wrought as a result of climate change might somehow be erased via the pounding feet of herds of fat, hairy elephants. If so, it's a bad scientific solution; but in the meantime, it's a great story.

■ ■ ■

By the early 1970s, with thousands abandoning the rural communes they once fled cities to build, the New Communalist vision was fading fast. So too were Brand's former ideals about routes to self-sufficiency. Instead, in the decade after he published the *Last Whole Earth Catalog* in 1971, Brand shifted the utopian rhetoric of personal empowerment he and others had previously placed on rural commune dwelling to a new focus: digital technology. According to Turner, this shift "helped redefine the microcomputer as a 'personal' machine, computer communication networks as 'virtual communities,' and cyberspace itself as the digital equivalent of the western landscape into which so many communards set forth in the late 1960s: the 'electronic frontier.'"

Will de-extinction ever become a reality? Who knows, but it's not looking likely. Instead, like the New Communalists' visions of small-scale utopias contained within the *Whole Earth Catalog*, it's likely going to remain only as a relic of a greater shift. Looking back, its main quality as a story may be that it perfectly embodied the contemporary conversation on the power of biotechnology—overblown in both its hopes and its anxieties. We're reinventing nature, dudes. Because, *Jurassic Park*. ■

Azeen Ghorayshi is a science writer and a founding editor of Method Quarterly.

Drawn from Nature by A. Wilson.   Engraved by W. B.
1. *Passenger Pigeon.* 2. *Blue-mountain Warbler.* 3. *Hemlock W.*